T0134678

Advances in Intelligent Systems and Computing

Volume 936

The series "Advances in Intelligent Systems and Computing" contains publications on theory, applications, and design methods of Intelligent Systems and Intelligent Computing. Virtually all disciplines such as engineering, natural sciences, computer and information science, ICT, economics, business, e-commerce, environment, healthcare, life science are covered. The list of topics spans all the areas of modern intelligent systems and computing such as: computational intelligence, soft computing including neural networks, fuzzy systems, evolutionary computing and the fusion of these paradigms, social intelligence, ambient intelligence, computational neuroscience, artificial life, virtual worlds and society, cognitive science and systems, Perception and Vision, DNA and immune based systems, self-organizing and adaptive systems, e-Learning and teaching, human-centered and human-centric computing, recommender systems, intelligent control, robotics and mechatronics including human-machine teaming, knowledge-based paradigms, learning paradigms, machine ethics, intelligent data analysis, knowledge management, intelligent agents, intelligent decision making and support, intelligent network security, trust management, interactive entertainment, Web intelligence and multimedia.

The publications within "Advances in Intelligent Systems and Computing" are primarily proceedings of important conferences, symposia and congresses. They cover significant recent developments in the field, both of a foundational and applicable character. An important characteristic feature of the series is the short publication time and world-wide distribution. This permits a rapid and broad dissemination of research results.

** Indexing: The books of this series are submitted to ISI Proceedings, EI-Compendex, DBLP, SCOPUS, Google Scholar and Springerlink **

More information about this series at http://www.springer.com/series/11156

Pongsarun Boonyopakorn ·
Phayung Meesad · Sunantha Sodsee ·
Herwig Unger
Editors

Recent Advances in Information and Communication Technology 2019

Proceedings of the 15th International Conference on Computing and Information Technology (IC2IT 2019)

 Springer

Editors
Pongsarun Boonyopakorn
Faculty of Information Technology
King Mongkut's University of Technology
North Bangkok
Bangkok, Thailand

Phayung Meesad
Faculty of Information Technology
King Mongkut's University of Technology
North Bangkok
Bangkok, Thailand

Sunantha Sodsee
Faculty of Information Technology
King Mongkut's University of Technology
North Bangkok
Bangkok, Thailand

Herwig Unger
LG Kommunikationsnetze
FernUniversität in Hagen
Hagen, Germany

ISSN 2194-5357   ISSN 2194-5365   (electronic)
Advances in Intelligent Systems and Computing
ISBN 978-3-030-19860-2   ISBN 978-3-030-19861-9   (eBook)
https://doi.org/10.1007/978-3-030-19861-9

This Springer imprint is published by the registered company Springer Nature Switzerland AG
The registered company address is: Gewerbestrasse 11, 6330 Cham, Switzerland

Preface

Within the last few years, computer scientists and engineers as well as users had to face rough and fast changes in computer interfaces and of the abilities of machines and services offered. And this is only the beginning: the research results obtained let all of us expect more significant progress and challenges in our life within the next decade. A major contribution to these developments was brought about by achievements in the area of artificial intelligence. Machine learning, natural language processing, speech recognition, image and video processing are only the major research and engineering directions, which made possible autonomous driving, language assistants, automatic translation and answering systems as well as other innovative applications such as more human-oriented interfaces. Those changes also reflect economic changes in the world, which are increasingly dominated by the needs of an enhanced globalisation and worldwide cooperation (including its competitive aspects) and by emerging global problems.

All those developments were mostly carried out by private companies, dominated by a few great players in the Internet business. Universities very often just follow such fast developments instead of being the motors of innovation and, thus, may have to justify their existence within the next few years. Consequently, universities have to rethink their approach to teaching and research and return to be really innovative in the area of basic research. This includes to abolish the perverse and inefficient 'publish or perish' approach as well as the omnipresent practice of evaluating 'all and everything' and the overwhelming paperwork and to return to the roots of science, viz. the competition for the best strategic ideas within a dispute far ahead of what is presently needed, but what will be the challenge in future.

This also concerns scientific conferences, which became more and more inefficient market places to present, advertise and 'sell' one's latest results and to put money in the pockets of many commercial organisers, many of which are just like 'day flies' and disappear after some very few events, only.

Differing from those events, the International Conference on Computing and Information Technology (IC2IT) celebrates its 15th anniversary in 2019—in the year of its major sponsor's 60th birthday, the King Mongkut's University of Technology North Bangkok—by still obeying the classic, less stagy rules of science. To ensure high quality, 59 submissions from 19 countries were thoroughly reviewed by at least two, usually even three members of the International Programme Committee (IPC). Through a positive vote of at least two IPC members, 32 submissions have been accepted for the presentation at the conference and inclusion in the conference proceedings, which are published in the well-established and worldwide-distributed series on Advances in Intelligent and Soft Computing edited by Janusz Kacprzyk for seven years now.

In contrast to other conferences, in this volume the above addressed, typical areas of artificial intelligence such as data mining with its theory and applications, image and natural language processing as well as the needed computational environments are not presented in an isolated manner neglecting the complexity of contemporary computer systems. Reacting to user needs in real time mostly requires strong computing facilities and background knowledge, which usually would overload a single, in particular mobile device. Appropriate achievements in networking, mobile and cloud computing as well as computer organisation are incorporated in powerful, innovative solutions. Finally, software development and security methodologies must generate holistic, optimised solutions improving the users' comfort when applying new artefacts.

Naturally, to unite these different experiences and approaches in one conference and one book requires a great preparation effort. Therefore, the volume's editors have to thank the authors as well as the members of the programme committee and all participating university partners in both Thailand and overseas for their outstanding support and academic cooperation. As usual, a special 'thank you' is given to all staff members of the Faculty of Information Technology at King Mongkut's University of Technology North Bangkok, who have carried out many technical and organisational tasks.

For many years, the signees conducted and significantly influenced the organisation of the IC2IT conference. They think that it is now time to say 'goodbye' and put the organisation of IC2IT 2020 in the hands of younger colleagues, which already participated in the organisation of the 2019 event. They are eager to realise their ideas and bring the needed, new impulses to our event.

Finally, we are convinced that all authors, presenters and participants will make IC2IT 2019 a successful event. So we wish all of them a pleasant time in Bangkok and hope that the conference will find the right way for the next generation of scientists: fruitful discussions of competitive ideas, which sustainable new systems may incorporate.

July 2019   Phayung Meesad
                                                                                        Herwig Unger

Organisation

Program Committee

M. Ahmed	Waikato, New Zealand
M. Aiello	Uni Stuttgart, Germany
S. Auwatanamongkol	NIDA, Thailand
T. Bernard	Li-Parad/Syscom CReSTIC
S. Boonkrong	SUT, Thailand
S. Butcharoen	TOT, Thailand
M. Caspar	DLR, Germany
T. Chintakovid	KMUTNB, Thailand
K. Chochiang	PSU, Thailand
K. Dittakan	PSU, Thailand
T. Eggendorfer	HS Weingarten, Germany
M. Hagan	Okstate U, USA
W. Halang	FernUni, Germany
C. Haruechaiyasak	NECTEC, Thailand
W. Jitsakul	KMUTNB, Thailand
S. Jungjit	Thaksin U, Thailand
M. Komkhao	RMUTP, Thailand
P. Kropf	UniNE, Switzerland
M. Kubek	FernUni, Germany
K. Kyamakya	AAU, Austria
U. Lechner	UniBW, Germany
H. Lefmann	TU Chemnitz, Germany
Z. Li	FernUni, Germany
K. Nimkerdphol	RMUTT, Thailand
K. Pasupa	KMITL, Thailand
M. Phongpaibul	TU, Thailand
J. Polpinij	MSU, Thailand
N. Porrawatpreyakorn	KMUTNB, Thailand

M. Sodanil	KMUTNB, Thailand
S. Sodsee	KMUTNB, Thailand
T. Srikacha	TOT, Thailand
T. Sucontphunt	NIDA, Thailand
C. Thaenchaikun	PSU, Thailand
J. Thaenthong	PSU, Thailand
D. Thammasiri	NPRU, Thailand
K. Thongglin	PSU, Thailand
J. Thongkam	MSU, Thailand
N. Tongtep	PSU, Thailand
D. H. Tran	HNUE, Vietnam
D. Tutsch	Uni Wuppertal, Germany
M. Weiser	Okstate U, USA
K. Woraratpanya	KMITL, Thailand

Organising Partners

In cooperation with

King Mongkut's University of Technology North Bangkok (KMUTNB)
FernUniversitaet in Hagen, Germany (FernUni)
Chemnitz University, Germany (CUT)
Oklahoma State University, USA (OSU)
Edith Cowan University, Western Australia (ECU)
Hanoi National University of Education, Vietnam (HNUE)
Gesellschaft für Informatik (GI)
Mahasarakham University (MSU)
Ubon Ratchathani University (UBU)
Kanchanaburi Rajabhat University (KRU)
Nakhon Pathom Rajabhat University (NPRU)
Phetchaburi Rajabhat University (PBRU)
Rajamangala University of Technology Krungthep (RMUTK)
Rajamangala University of Technology Thanyaburi (RMUTT)
Prince of Songkla University, Phuket Campus (PSU)
National Institute of Development Administration (NIDA)
Sisaket Rajabhat University (SSKRU)
Council of IT Deans of Thailand (CITT)
IEEE CIS Thailand

Contents

Data Mining

**Dual Increment Shapelets: A Scalable Shapelet Discovery
for Time Series Classification** 3
Nattakit Vichit and Chotirat Ann Ratanamahatana

**Analyzing and Visualizing Anomalies and Events in Time Series
of Network Traffic** 15
Qinpei Zhao, Yinjia Zhang, Yang Shi, and Jiangfeng Li

**The Grid-Based Spatial ARIMA Model: An Innovation
for Short-Term Predictions of Ocean Current Patterns
with Big HF Radar Data** 26
Ratchanont Pongto, Nopparat Wiwattanaphon, Peerapon Lekpong,
Siam Lawawirojwong, Siwapon Srisonphan, Kerk F. Kee,
and Kulsawasd Jitkajornwanich

**Parameter-Free Outlier Scoring Algorithm Using the Acute Angle
Order Difference Distance** 37
Pollaton Pumruckthum, Somjai Boonsiri, and Krung Sinapiromsaran

**Improved Weighted Least Square Radiometric Calibration Based
Noise and Outlier Rejection by Adjacent Comparagraph
and Brightness Transfer Function** 46
Chanchai Techawatcharapaikul, Pradit Mittrapiyanurak,
and Werapon Chiracharit

Application of Data Mining, Language, and Text Processing

**Fuzzy TF-IDF Weighting in Synonym for Diabetes Question
and Answers** 59
Ketsara Phetkrachang and Nichnan Kittiphattanabawon

Feature Comparison for Automatic Bug Report Classification 69
Bancha Luaphol, Boonchoo Srikudkao, Tontrakant Kachai,
Natthakit Srikanjanapert, Jantima Polpinij, and Poramin Bheganan

A Novel Three Phase Approach for Single Sample Ear Recognition . . . 79
Nitin Kumar

**Comparison of Thai Sentence Sentiment Tagging Methods
Using Thai Sentiment Resource** 89
Kanlaya Thong-iad and Ponrudee Netisopakul

**Collecting Child Psychiatry Documents of Clinical Trials
from PubMed by the SVM Text Classification Method
with the MATF Weighting Scheme** 99
Jantima Polpinij, Tontrakant Kachai, Kanyarat Nasomboon,
and Poramin Bheganan

Image Processing

**Instance-Based Learning for Blood Vessel Segmentation
in Retinal Images** 111
Worapan Kusakunniran, Sarattha Kanchanapreechakorn,
and Kittikhun Thongkanchorn

**Floor Projection Type Serious Game System for Lower Limb
Rehabilitation Using Image Processing** 119
Kazuo Hemmi, Yuki Kondo, Takuro Tobina, and Takeshi Nishimura

**Image Processing Technique for Gender Determination
from Medical Microscope Image** 129
Wichan Thumthong, Hathaichanok Chompoopuen, and Pita Jaupunphol

**Accelerate the Detection Frame Rate of YOLO
Object Detection Algorithm** 138
Wattanapong Kurdthongmee

**Analyze Facial Expression Recognition Based on Curvelet
Transform via Extreme Learning Machine** 148
Sarutte Atsawaruangsuk, Tatpong Katanyukul, and Pattarawit Polpinit

**Ensemble Model for Segmentation of Lateral Ventricles from 3D
Magnetic Resonance Imaging** 159
Akadej Udomchaiporn, Khitichai Lertrungwichean, Pokpakorn Klinkasen,
and Chawanwut Nuchprasert

**Deep Convolutional Neural Network with Edge Feature
for Image Denoising** 169
Supakorn Chupraphawan and Chotirat Ann Ratanamahatana

The Combination of Different Cell Sizes of HOG with KELM
for Vehicle Detection 180
Natthariya Laopracha

Network, Cloud and Management

Dynamic Data Management for an Associative P2P Memory 193
Supaporn Simcharoen

Extremely Fast Neural Computation Using Tally
Numeral Arithmetic 205
Kosuke Imamura

Traceable CP-ABE for Outsourced Big Data in Cloud Storage 213
Praveen Kumar Premkamal, Syam Kumar Pasupuleti,
and P. J. A. Alphonse

A Novel Solution for Virtual Server on the Data Consistency
Maintenance in Cloud Storage Systems 227
Van Thang Doan, Vo Quang Hoang Khang, Ha Huy Cuong Nguyen,
Cong Phap Huynh, and Phayung Meesad

Data Integration Patterns in the Context of Enterprise
Data Management 235
Roland Petrasch

Evolutionary Dynamics of Service Provider Legacy Network
Migration to Software Defined IPv6 Network 245
Babu R. Dawadi, Danda B. Rawat, and Shashidhar R. Joshi

Hiding Patient Injury Information in Medical Images
with QR Code 258
Akkarat Boonyapalanant, Mahasak Ketcham,
and Manussawee Piyaneeranart

Author Index. 269

Data Mining

Dual Increment Shapelets: A Scalable Shapelet Discovery for Time Series Classification

Nattakit Vichit and Chotirat Ann Ratanamahatana[✉]

Department of Computer Engineering, Faculty of Engineering,
Chulalongkorn University, 254 Phayathai Road, Pathumwan,
Bangkok 10330, Thailand
6070178221@student.chula.ac.th,
chotirat.r@chula.ac.th

Abstract. As time series data become more complex and users expect more sophisticated information, numerous algorithms have been proposed to solve these challenges. Among those algorithms to classify time series data, shapelet – a discriminative subsequence of time series data – is considered a practical approach due to its accurate and insightful classification. However, previously proposed shapelet algorithms still suffer from exceedingly high computational complexity, as a result, limiting its scalability to larger datasets. Therefore, in this work, we propose a novel algorithm that speeds up shapelet discovery process. Our algorithm so called "Dual Increment Shapelets (DIS)" is a combination of two-layered incremental neural network and filtering process based on subsequence characteristics. Empirical experiments on forty datasets evidently demonstrate that our proposed work could achieve large speedup while maintaining its accuracy. Unlike the previous algorithm that mainly emphasizes speedup of the search algorithm, DIS essentially reduces the number of shapelet candidates based on subsequence characteristics. As a result, our DIS algorithm could achieve more than three orders of magnitude speedup, comparing with the baseline algorithms, while preserving the accuracy of the state-of-the-art algorithm.

Keywords: Time series · Data mining · Classification · Shapelet

1 Introduction

A time series is a sequence of data point which represents data in a time interval. Apart from classical time series data such as stock market data, other types of data have recently been shown to work effectively and efficiently for various data mining tasks once transformed into time series data, including shape classification, movement tracking, medical diagnosis (ECG/EKG, EEG, etc.), motif discovery, anomaly detection, classification, clustering, etc.

In working with time series data, many challenges that affect the performance of the time series mining tasks include occlusion, distortion of warping, uniform scaling, uncertainty, and wandering baselines [1]. Therefore, the algorithms must be designed to overcome the complexity of a time series with such invariances.

© Springer Nature Switzerland AG 2020
P. Boonyopakorn et al. (Eds.): IC2IT 2019, AISC 936, pp. 3–14, 2020.
https://doi.org/10.1007/978-3-030-19861-9_1

A large amount of studies have been conducted to improve time series classification tasks. A well known approach is based on distance, such as Euclidean Distance (ED) or Dynamic Time Warping (DTW) combined with 1-NN [2]. However, these algorithms have been shown to be sensitive to missing data that might lead to misclassification. To solve this problem, many researchers opted to classify time series data by discovering and using local features rather than the whole time series sequence, e.g., interval-based classifiers, dictionary-based classifiers, and shapelet-based classifiers.

Above all three aforementioned algorithms, the shapelet-based classifiers have been recently shown to provide the most accurate and interpretable results with the speediest testing. Due to its flexibility, the shapelet-based classifiers are applicable for both local features and whole time series. This flexible quality allows the extensive improvement of the classification accuracy. Comparing with the lazy algorithm such as DWT based 1-NN classifier, the eager algorithm like shapelet-based classifiers could save more time to classify the target classes. Moreover, shapelet-based classifiers provide interpretable results, which facilitate the domain experts in the fields to effectively interpret the results. For example, the shapelets that are generated by the shapelet-based classifiers could evidently provide prominent features of 'Urtica Dioica' and 'Verbena Urticifolia' [3]. Based on these shapelet-based classifier results, the botanists could identify the types of the leaves by having only a quick glance of the shapelets.

Another application of shapelet-based classifiers are classification of video clips of actions; a hand's centroid position in each frame is tracked. Figure 1 compares time series sequences of a 150-frame video clip between the actions of a person grasping a gun from a holster, pointing it to a target, then putting it back to the holster versus another person resting his hand on the side, only pointing his finger to a target, then putting his hand back to the resting position. As seen in the figure, the 100th–130th time interval clearly shows the difference between the person holding a gun and the one without. Even though the two time series sequences demonstrate the discrepancies between the two actions in the 50th–90th time interval (different heights of the hand pointing a finger vs. pointing a gun), the shapelet-based classifiers see them as irrelevant invariances and successfully differentiate the two actions using only the essential intervals. It can be observed that the overly captured invariances could mislead the results. Thus, shapelet-based classifiers are considered more appropriate and effective.

Fig. 1. A comparison between time series sequences of the actions of a person holding and pointing a gun (Gun time series) and a person pointing a finger (No gun time series).

In spite of its benefits, shapelet-based classifiers still suffer from a major drawback; they are too slow that they become infeasible with large datasets. Even in state-of-the-art approach, the algorithm still requires to generate a large number of shapelet candidates. A group of shapelet candidates could be as large as $O(n^2m^4)$ when n is the number of time series in the dataset, and m is the length of each time series. For instance, if a dataset contains only 70 instances, where each instance is 500 data points long, the shapelet candidates could accumulate to more than 10^{12} candidates, which become infeasible for the classifier to give prediction results within reasonable amount of time unless adequate number of candidates are pruned out.

To resolve the issues of infeasibility and practicality of the existing shapelet-based classifiers, we design a new algorithm, which optimizes the shapelet candidate selection process, while preserving the accuracy of the results and no-false-dismissal property. More specifically, our proposed work can effectively prune out shapelet candidates using a so-called "Dual Increment Shapelets (DIS)" that is a combination of two-layered incremental neural network and filtering process based on subsequence characteristics.

The rest of the paper is organized as follows. Section 2 gives fundamental background on shapelets and reviews on its related work. Section 3 presents our proposed algorithm. Subsequently, the empirical evidence, interpretation of the results, and discussions are provided in Sect. 4. Finally, the conclusion, limitations, and suggestions for future studies are discussed in Sect. 5.

2 Related Work

2.1 Shapelet

Shapelet is introduced by [4] as a decision tree algorithm to search for the best shapelet candidates. Shapelet candidates are generated through testing on every possible subsequence length. All shapelet candidates will then be evaluated on their qualities using an information gain measure, which returns the best split point. For example, as shown in Fig. 2, a group of data in the first half (represented by circles) has better quality than a group of data in the second half (represented by a mixture of rectangles and a circle).

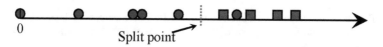

Fig. 2. A split point is determined to provide the best quality within each group of the data [4]

Since the traditional process of finding shapelets is very time-consuming, Fast shapelet [3] has been introduced to reduce the running time through dimensionality reduction (SAX) [5] and random projection [6] to group similar shapelets. Regardless of an order of magnitude speedup, the accuracy has to be significantly sacrificed.

Shapelet Transform is then proposed by [5]. Its shapelet finding process is separated from the classifier, which allows us to use any classifiers and shapelet quality measures. Its running time could be improved by opting in faster shapelet quality evaluation such as Mood's Median and F-stat [7]. By using F-stat quality evaluation, more than two orders of magnitude speedup can be achieved because it removes the need of checking every node in the subtrees. To improve the accuracy, inferior decision tree classifiers could be replaced with better algorithms such as Rotation Forest [8] or Support Vector Machine [9].

Learning Shapelet are introduced by [10]. Instead of looping and evaluating the accuracy of shapelet candidates one by one, learning shapelet uses a gradient descent to learn near-optimal shapelets.

To compare among these baseline algorithms, both accuracy and speed are desirable. In terms of accuracy, Shapelet Transform is the most accurate as referred to the as state of the art, followed by Learning Shapelet and Fast Shapelet, respectively. On the contrary, as expected, in terms of speed, Shapelet Transform is the slowest, while Fast Shapelet is the fastest.

2.2   Shapelet Characteristic Filtering

As shapelet candidates vary in their shapes and characteristics, we can use these attributes to filter out useless candidates. In fact, key points of shapelets can be selected using this criteria [11]; the points that most reduce the distance to original time series are selected to generate shapelet candidates.

As seen in Fig. 3, selected points (KP3-KP4 and KP5-KP6-KP7) are used to generate shapelet candidates. With this strategy, we do not only reduce a lot of data points and shapelet candidates, but also maintain the best-quality candidates.

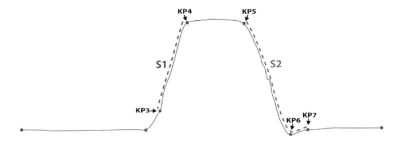

Fig. 3. Examples of shapelet candidates generated by key points. Candidate S1 is generated from key point 3 (KP3) to key point 4 (KP4), and candidate (S2) is generated from key point 5 (KP5) to key point 7 (KP7)

2.3   Self-organizing Incremental Neural Network

Self-organizing incremental neural network is introduce by [12]. It uses DTW combined with a neural network to cluster the data in one pass. The amount of an output group is varied by the size of the dataset. Incremental shapelet [13] acts as a shapelet candidate selector. First, it generates shapelets by using the skip length parameter in shapelet length. Then, it employs self-organizing incremental neural network method to select shapelet candidate. However, this skip length parameter can significantly reduce classification accuracy [14]. Another downside is that the incremental shapelet also limits the same shapelet candidate length to be grouped together.

3 Our Proposed Work: Dual Increment Shapelets

Our algorithm is divided into 3 steps as follows:

3.1 Shapelet Candidate Selection

Not all shapelet candidates have the same quality. A Candidate with large variation is considered having higher quality than a candidate with small variation (e.g., straight line). We adopt a Local Farthest Deviation Points (LFDP) algorithm [15] to select only candidates with high variation. For each time series sequence, LFDP algorithm first marks the start and end of the set of selected points and then set the distance to infinity (lines 2–4). LFDP algorithm continues selecting a point until its distance to the original time series falls below the threshold (lines 5–8). The selected points are all appended together to the list of important points (line 9). These reference points are used to mark the starting and ending points of a shapelet candidate, as shown in Fig. 3. As a result, we could filter out a large number of shapelet candidates, while maintaining shapelet candidate's quality.

Algorithm 1 LFDP [12]

1: **function** LFDP($timeSeries$)
2:     $importantPointList \leftarrow [\ ]$
3:     **for all** $timeSeries$ **do**
4:         $importantPoints \leftarrow [start, end]$
5:         $distance \leftarrow infinity$
6:         **while** $distance > threshold$ **do**
7:             $tempTimeSeries \leftarrow$ createTimeSeries($importantPoints$)
8:             $distance \leftarrow$ distance($tempTimeSeries, timeSeries$)
9:             $importantPoints \leftarrow$ findNextBestCandidatePoint($timeSeries$)
10:        $importantPointList$.append($importantPoints$)
        return $importantPointList$

3.2 Incremental Neural Network

An Incremental Neural Network is our preferable choice of clustering algorithm as it is a one-pass algorithm, which is suitable for large datasets. Moreover, it has an ability to self-adjust the number of groups by candidate's variation. We adopt this idea from work in [11]. However, instead of one single layer, we have modified the algorithm to become a two-layered network, i.e., Candidates Averaging layer, and Class Purity layer.

Candidates Averaging Layer

A large number of candidates are similar in shape and size. Similar candidates would result in similar shapelet quality. If we cluster similar shapelets together, we would be able to estimate the quality of shapelets groupwise rather than individually. We use the length and shapelet average within the same group to calculate shapelet quality.

The first layer will average candidates of the same length from the same original time series sequence, whose distance between the candidates was below the threshold. As shown in Algorithm 2, the inputs are candidates that are grouped by their length. For all candidates with the same length (line 4), we insert them one by one into our incremental neural network (line 5). After inserting all data with the same length, we then average of all candidates in each node and append it to the list of filtered candidates (line 6). The algorithm continues until all lengthwise candidate grouping are completed (line 2). The outputs are the averaged filtered candidates, based on their lengths.

Algorithm 2 Candidate Averaging Layer

1: **function** CANDIDATEAVERAGINGLAYER(tuples($length, candidates$))
2:     $filteredCandidates \leftarrow [\,]$
3:     **for all** $length$ get $candidates$ **do**
4:         $neuralNetwork \leftarrow$ create IncrementalNetwork()
5:         **for all** $candidates$.get($length$) **do**
6:             $neuralNetwork.insertNewData(candidate)$
7:         $filteredCandidates \leftarrow neuralNetwork$.average()
        return $filteredCandidates$

Class Purity Layer

The measurement criteria for the best shapelet relies on the power of class discrimination [16]. To measure shapelet quality, we adopt the idea from [3] and [16] in measuring the quality of the group using class purity of a shapelets group. During the shapelet grouping process, we insert the class metadata into the groups to help in group purity calculation. We used this approximated group quality to filter out some low-quality candidate groups instead of selecting each candidate one by one. Regardless of the length, this second layer could effectively group similar shapelets together.

Algorithm 3 Class Purity Layer

1: **function** CLASSPURITYLAYER($candidates$)
2:     $neuralNetwork \leftarrow$ create IncrementalNetwork()
3:     **for all** $candidate$ in $candidates$ **do**
4:         $neuralNetwork$.insertNewData($candidate$)
5:     $filteredCandidates \leftarrow neuralNetwork$.quality()
6:     **return** $filteredCandidates$

3.3   Dual Increment Shapelets

With those strategies above, we could actually reduce the number of candidates to be evaluated by a few orders of magnitude. The remaining groups have good candidate potentials. All of the remaining candidates are then used to find the distance to all time series in the dataset. The underlying algorithm is very similar to the shapelet transform. Lastly, the results of the transform step are classified by SVM algorithm with a purpose to find the best shapelet candidates. With L1 regularization [17], weights are consolidated with some candidates that contribute the best accuracy. As a result, an acceptable number of shapelets are acquired.

Algorithm 4 Dual Increment Shapelets

1: **function** DUALINCREMENTSHAPELETS($timeSeries$)
2:     $importantPoints \leftarrow$ LFDP(timeSeries)
3:     $shapeletCandidates \leftarrow$ candidateGeneration($importantPoints$)
4:     $tuples(length, candidates) \leftarrow$ sortByLength($shapeletCandidates$)
5:     $averageShapelets \leftarrow$ candidateAveragingLayer($tuples(length, candidates)$)
6:     $(shapeletGroup, quality) \leftarrow$ classPurityLayer($averageShapelets$)
7:     $selectedCandidates \leftarrow [\ ]$
8:     **for all** $(shapeletsGroup, quality)$ **do**
9:         **if** $quality > threshold$ **then**
10:             $selectedCandidates \leftarrow shapeletGroup$
11:     $shapelets \leftarrow$ SVM-L1reg($selectedCandidates$)
12:     **return** $shapelets$

The dual increment shapelets algorithm starts by first selecting important points obtained from the LFDP algorithm. Then, those selected points are used to generate shapelet candidates (lines 1–2). The generated candidates were transferred to the first layer of the incremental neural network to average candidates that share the same length (line 3). After that, these candidates were transferred to the second layer of the incremental neural network which hosted a group of similar shapelet candidates (line 4). The quality of every group in layer 2 is measured by the class purity (lines 6–7). The selected groups above the threshold are kept in the selected candidate list (line 8). Then, they are classified by SVM with L1 regularization to obtain the best shapelets (line 9).

4   Experiments and Results

To examine the accuracy and effectiveness of our proposed DIS, the datasets from the UCR repository [18] are tested across all four shapelet-based algorithms: our proposed DIS, Shapelet Transform (ST), Learning Shapelet (LS), and Fast Shapelet (FS). In this paper, the three algorithms (ST, LS, and FS) will be referred to as the state-of-the-art baselines. Using Java WEKA framework [19], we acquire the codes from [10] to reimplement and recreate the results of the baselines. However, due to the exceedingly large time complexity, some baseline algorithms are unable to complete the classification in some large datasets within 24-h period. Specifically, only 40 datasets from the total of 76 datasets could be run by Shapelet Transform algorithm. Therefore, even though our proposed DIS algorithm could be successfully trained within the time limit, for fair comparison, only these 40 datasets are chosen for the experiments.

4.1 Running Time Comparison to the Baselines

To evaluate the performance, every algorithm is repeated 5 times, and the average running time is reported. Table 1 reports the average running time for all shapelet-based classifiers.

Table 1. Average running time in millisecond for shapelet-based classifiers

Datasets	FS	LS	ST	DIS
Adiac	101177	99454633	10692757	**14834**
ArrowHead	10815	228300	4841654	**1596**
Beef	75734	1915378	3123344	**9364**
BeetleFly	22934	179908	5866064	**9474**
BirdChicken	17178	159999	6281603	**8060**
Car	119352	3926827	50858285	**52557**
CBF	3301	49966	213188	**516**
ChlorineConcentration	191170	1820407	52458469	**12248**
Coffee	6244	76629	302256	**1013**
DiatomSizeReduction	6882	309793	93089	**831**
DistalPhalanxOutlineAgeGroup	10385	410264	15641115	**2609**
DistalPhalanxOutlineCorrect	23501	287386	10530021	**3377**
DistalPhalanxTW	13456	1933645	5874907	**2045**
ECGFiveDays	2267	15433	159845	**224**
FaceAll	190672	37933408	78029678	**26308**
FaceFour	24148	500485	7792586	**1994**
FacesUCR	54225	12490998	19879931	**5122**
GunPoint	2578	45133	895632	**679**
ItalyPowerDemand	1166	4777	2730	**295**
Lightning7	87039	4702008	54935741	**19362**
MedicalImages	46871	7818573	28067449	**6739**
MiddlePhalanxOutlineAgeGroup	10184	412895	8862617	**1526**
MiddlePhalanxTW	14331	1917648	14721367	**3744**
MoteStrain	1738	6538	10036	**175**
OliveOil	47847	1623811	3082948	**8836**
PhalangesOutlinesCorrect	88116	968687	99125785	**16083**
Plane	14181	1658518	10914489	**2583**
ProximalPhalanxOutlineAgeGroup	9819	410964	8685065	**1355**
ProximalPhalanxOutlineCorrect	18283	269331	7089362	**2377**

(*continued*)

Table 1. (*continued*)

Datasets	FS	LS	ST	**DIS**
ProximalPhalanxTW	10817	1938259	9705892	**1767**
ShapeletSim	31187	210767	542623	**7698**
SonyAIBORobotSurface1	1504	4862	8410	**220**
SonyAIBORobotSurface2	1561	6323	12257	**281**
SwedishLeaf	104532	36721071	86573650	**16996**
Symbols	24673	1648839	15207771	**4569**
SyntheticControl	10487	934441	2413403	**3405**
ToeSegmentation1	11377	109016	8114483	**3792**
ToeSegmentation2	14778	197764	17057920	**6911**
Trace	45370	1624364	87329377	**15214**
TwoLeadECG	1468	7206	4381	**175**

Table 1 shows that our proposed DIS algorithm evidently and significantly out-performs all other algorithms in terms of the running time. Especially in multiclass problems, DIS has an extra ability to prune more candidates than the binary class problems, being as much as 6,400 times faster in some datasets. These results indicate that the characteristics of the datasets (length, number of instances, number of classes, etc.) could be influential factors affecting the speed of DIS. In the other words, the characteristics of the dataset directly affect the pruning power of the shapelet candidates, e.g., ItalyPowerDemand vs. FaceAll.

The types of algorithms also influence the running time. The results from Table 1 revealed that our DIS outperformed all the baselines since it is a one-pass candidate filtering algorithm. To be more specific, comparing to DIS, Shapelet Transform is the slowest one (2460.567 times slower), followed by Learning Shapelet (525.664 times slower), and Fast Shapelet (6.180 times slower), respectively, as shown in Table 2. It can be explained that Brute Force algorithm like Shapelet Transform requires very large running time to run. The Gradient descent algorithm underlying the Learning Shapelet is faster than the Brute Force algorithm, and the dimensionality reduced algorithm could greatly help reduce the running time, making Fast Shapelet the fastest one among all the baselines.

Table 2. A comparison of the results of average speedup of DIS and the baselines

Fast shapelet	Learning shapelet	Shapelet transform
6.180	525.664	2460.567

4.2   Accuracy Comparison to the Baselines

In terms of accuracy, our proposed DIS and the baselines were compared in order to compare the accuracy of DIS against all the baselines. It should be noted that the Shapelet Transform employed in this session was parameterized using the same parameters reported in [20]. The results are provided in Table 3.

Table 3. Classification accuracies for shapelet-based classifiers

Dataset	FS	LS	ST	**DIS**
Adiac	0.550	0.519	0.130	**0.729**
ArrowHead	0.577	**0.823**	0.720	0.777
Beef	0.567	**0.800**	0.567	0.767
BeetleFly	0.650	0.750	**0.800**	**0.800**
BirdChicken	**0.900**	0.800	0.750	**0.900**
Car	0.733	0.800	0.633	**0.883**
CBF	0.919	**0.990**	0.956	0.941
ChlorineConcentration	0.591	0.591	0.616	**0.648**
Coffee	0.964	**1.000**	**1.000**	0.964
DiatomSizeReduction	0.879	**0.967**	0.765	0.941
DistalPhalanxOutlineAgeGroup	0.640	**0.719**	0.691	0.698
DistalPhalanxOutlineCorrect	0.728	**0.786**	0.670	0.775
DistalPhalanxTW	**0.655**	0.626	0.647	0.647
ECGFiveDays	0.995	**1.000**	**1.000**	**1.000**
FaceAll	0.620	0.775	0.653	**0.778**
FaceFour	0.920	**0.966**	0.750	0.943
FacesUCR	0.738	**0.944**	0.671	0.878
GunPoint	0.940	**1.000**	0.953	0.993
ItalyPowerDemand	0.906	**0.963**	0.943	0.958
Lightning7	0.630	**0.808**	0.425	0.740
MedicalImages	0.605	0.686	0.471	**0.689**
MiddlePhalanxOutlineAgeGroup	0.539	**0.584**	0.532	**0.584**
MiddlePhalanxTW	0.461	0.506	**0.526**	0.494
MoteStrain	0.798	0.858	0.839	**0.874**
OliveOil	0.633	0.700	0.767	**0.867**
PhalangesOutlinesCorrect	0.724	0.748	0.685	**0.833**
Plane	0.990	**1.000**	0.924	**1.000**
ProximalPhalanxOutlineAgeGroup	0.776	0.815	0.737	**0.834**
ProximalPhalanxOutlineCorrect	0.838	0.849	0.715	**0.887**
ProximalPhalanxTW	0.727	**0.810**	0.654	0.790
ShapeletSim	**1.000**	0.978	**1.000**	0.972
SonyAIBORobotSurface1	0.686	0.827	**0.947**	0.429
SonyAIBORobotSurface2	0.790	**0.890**	0.876	0.824
SwedishLeaf	0.789	0.917	0.755	**0.928**
Symbols	**0.937**	0.930	0.823	0.930
SyntheticControl	0.937	**0.997**	0.957	0.990
ToeSegmentation1	0.943	0.930	0.934	**0.956**
ToeSegmentation2	0.692	**0.923**	0.892	0.800
Trace	**1.000**	**1.000**	0.980	**1.000**
TwoLeadECG	0.946	**0.997**	0.970	0.991

As the tradeoff for speed, our accuracies are expected to be lowered. However, DIS does surprisingly well, being a winner among all baselines in as many as 18 datasets. The Gradient descent algorithm in Learning Shapelet could yield good results as expected, but the Fast Shapelet is the least accurate despite its fastest speed, due to the dimensionality reduced data.

Table 4 reports the performance of our proposed DIS algorithm in terms of accuracy, comparing to the three baselines. Our DIS generally outperforms Fast Shapelet and Shapelet Transform, but is slightly less accurate than the Learning Shapelet. One possible explanation involves their underlying behavior of the algorithms. The Gradient descent algorithm could effectively search for the qualified shapelets. The DIS might prune some qualified shapelet candidates out, leading to the possible deviated results.

Table 4. A comparison of the results of accuracy of DIS and each baseline algorithm

Result	Fast shapelet	Learning shapelet	Shapelet transform
Win	33	15	30
Lose	4	20	7
Tie	3	5	3

5 Conclusions and Future Work

In this work, a novel algorithm so-called "Dual Increment Shapelets (DIS)" is introduced. We evaluate its performance in terms of speed and accuracy against the three baselines: Fast Shapelet, Learning Shapelet, and Shapelet Transform. The results reveal that our DIS is the fastest algorithm comparing with all the baselines. With the significantly improved speed, the accuracy of DIS has only been slightly sacrificed. The key to its success is good candidate filtering ability, which is a result of LFDP candidate generation method combined with two-layered Incremental Neural Network. With these capacities, our DIS offers a promising algorithm to handle large datasets while maintaining satisfactorily high accuracy. However, as a future work, we could improve its ability to pin-point qualified shapelets, as a result, speeding up the training time.

References

1. Keogh, E., Wei, L., Xi, X., Vlachos, M., Lee, S.-H., Protopapas, P.: Supporting exact indexing of arbitrarily rotated shapes and periodic time series under Euclidean and warping distance measures. VLDB J. **18**, 611–630 (2009)
2. Keogh, E., Ratanamahatana, C.A.: Exact indexing of dynamic time warping. Knowl. Inf. Syst. **7**, 358–386 (2005)

3. Rakthanmanon, T., Keogh, E.: Fast shapelets: a scalable algorithm for discovering time series shapelets. In: Proceedings of the 2013 SIAM International Conference on Data Mining, pp. 668–676. SIAM (2013)
4. Ye, L., Keogh, E.: Time series shapelets: a new primitive for data mining. In: Proceedings of the 15th ACM SIGKDD International Conference on Knowledge Discovery and Data Mining, pp. 947–956. ACM (2009)
5. Lin, J., Keogh, E., Wei, L., Lonardi, S.: Experiencing SAX: a novel symbolic representation of time series. Data Min. Knowl. Discov. **15**, 107–144 (2007)
6. Bingham, E., Mannila, H.: Random projection in dimensionality reduction: applications to image and text data. In: Proceedings of the Seventh ACM SIGKDD International Conference on Knowledge Discovery and Data Mining, pp. 245–250. ACM (2001)
7. Lines, J., Bagnall, A.: Alternative quality measures for time series shapelets. In: International Conference on Intelligent Data Engineering and Automated Learning, pp. 475–483. Springer (2012)
8. Rodriguez, J.J., Kuncheva, L.I., Alonso, C.J.: Rotation forest: a new classifier ensemble method. IEEE Trans. Pattern Anal. Mach. Intell. **28**, 1619–1630 (2006)
9. Rodríguez, J.J., Alonso, C.J.: Support vector machines of interval-based features for time series classification. In: International Conference on Innovative Techniques and Applications of Artificial Intelligence, pp. 244–257. Springer (2004)
10. Grabocka, J., Schilling, N., Wistuba, M., Schmidt-Thieme, L.: Learning time-series shapelets. In: Proceedings of the 20th ACM SIGKDD International Conference on Knowledge Discovery and Data Mining, pp. 392–401. ACM (2014)
11. Ji, C., Liu, S., Yang, C., Pan, L., Wu, L., Meng, X.: A shapelet selection algorithm for time series classification: new directions. Procedia Comput. Sci. **129**, 461–467 (2018)
12. Okada, S., Hasegawa, O.: Motion recognition based on dynamic-time warping method with self-organizing incremental neural network. In: 2008 19th International Conference on Pattern Recognition, ICPR 2008, pp. 1–4. IEEE (2008)
13. Yang, Y., Deng, Q., Shen, F., Zhao, J., Luo, C.: A shapelet learning method for time series classification. In: 2016 IEEE 28th International Conference on Tools with Artificial Intelligence (ICTAI), pp. 423–430. IEEE (2016)
14. Bostrom, A., Bagnall, A., Lines, J.: Evaluating Improvements to the Shapelet Transform (2016). www-bcf.usc.edu
15. Ji, C., Liu, S., Yang, C., Wu, L., Pan, L., Meng, X.: A piecewise linear representation method based on importance data points for time series data. In: 2016 IEEE 20th International Conference on Computer Supported Cooperative Work in Design (CSCWD), pp. 111–116. IEEE (2016)
16. Bostrom, A., Bagnall, A.: Binary shapelet transform for multiclass time series classification. Trans. Large-Scale Data Knowl.-Cent. Syst. **XXXII**, 24–46 (2017)
17. Zhu, J., Rosset, S., Tibshirani, R., Hastie, T.J.: 1-norm support vector machines. In: Advances in Neural Information Processing Systems, pp. 49–56 (2004)
18. Chen, Y., Keogh, E., Hu, B., Begum, N., Bagnall, A., Mueen, A., Batista, G.: The UCR time series classification archive, July (2015)
19. Hall, M., Frank, E., Holmes, G., Pfahringer, B., Reutemann, P., Witten, I.H.: The WEKA data mining software: an update. ACM SIGKDD Explor. Newsl. **11**, 10–18 (2009)
20. Bagnall, A., Lines, J., Bostrom, A., Large, J., Keogh, E.: The great time series classification bake off: a review and experimental evaluation of recent algorithmic advances. Data Min. Knowl. Discov. **31**, 606–660 (2017)

Analyzing and Visualizing Anomalies and Events in Time Series of Network Traffic

Qinpei Zhao, Yinjia Zhang, Yang Shi, and Jiangfeng Li[✉]

Tongji University, Shanghai, China
{shiyang,lijf}@tongji.edu.cn

Abstract. The traffic among the hosts and behaviors of the anomalous hosts in the network is usually complex. In network traffic, there is a key problem that is how to identify the security incidents. The corresponding question that who have contributed to the incidents is arisen then. A method, which detects both anomalies and events at the same time is quite helpful. A data from network traffic can be composed of the hosts and different attributes (traffic flow like amount of upload package and download package) in time series. Based on the structure of the network traffic data, we propose an anomaly and event detection method based on the network attributes in time series. The method analyzes both the host's behavior and the temporal features of the network traffic.

Keywords: Network traffic · Anomaly detection · Security incidents · Time series

1 Introduction

More and more devices are being linked to the Internet, there is gigabytes of traffic flowing every second through network devices. Therefore, the network security becomes imperative and network traffic anomaly detection constitutes an important part of the network security.

The network security is threaten by different attacks from malicious sources, such as Denial of service attack (DoS), large-volume point-to-point flows (*alpha flows*), worms, traffic from a single source to many destinations (*point-to-multipoint*) and so on [1]. The attacks usually show features of anomalies in a traffic of network. For a common network, a traffic flow with unusual and significant changes is considered as an anomaly. The network traffic has several kinds of attributes, for example traffic volumes, IP addresses and port numbers. The changes of the attributes, such as volume in traffic flows, distribution patterns of IP source and/or destination addresses and port numbers etc. are typically signs of anomalies. When there is a significantly changed traffic flow in the network, it could be identified as an anomaly.

The network traffic data can be obtained at multiple levels of granularity, such as data characterizing TCP/UDP or data gathered from network equipments (routers, switches). Time as an important factor among the network needs

© Springer Nature Switzerland AG 2020
P. Boonyopakorn et al. (Eds.): IC2IT 2019, AISC 936, pp. 15–25, 2020.
https://doi.org/10.1007/978-3-030-19861-9_2

to be included into the data [2–4]. Network traffic data with attributes of time stamps, source and destination IP addresses, port numbers and even text, is often related to each other and shows dependencies.

There have been a lot of research on the network traffic anomaly detection methods [5]. Several surveys are available in the literatures. A structured and comprehensive overview of various facets of network anomaly detection has been given in [6]. A large number of detection methods and systems, especially tools are compared in the paper. A tutorial focusing on statistical techniques (also data mining techniques) is presented in [1]. Three major methods of statistical techniques: PCA-based [7,8], sketch-based and signal-analysis-based, are compared. Signal processing techniques have been successfully applied to the problem due to their ability in point change detection and data transformation. Wavelets are excellent tool for computing aggregates at different scales [9,10]. Combining various anomaly detectors through consensus optimization is also a choice [11].

The authors in [12] believe that change in network traffic over time is well suited for detecting uncommon system behaviors. The importance of time information is enhanced in anomaly detection. Therefore, we first introduce the *S*liding *W*indow based *E*vent and *A*nomaly detection method for the network traffic in *T*ime series (*SWEAT*). The sliding window is employed for getting different resolutions of results. The novelty of the method is that not only the time series of each host is analyzed, but the correlation among the hosts is also considered. Therefore, the method is capable of detecting the event from time aspect, and locating the hosts contribute to the event.

2 Preliminaries

The traffic of a network is commonly composed of a set of hosts $P = \{p_1, p_2, ..., p_N\}$. Given a raw data of network traffic, the first step is to perform data cleaning and feature engineering. After the preprocessing on the raw data, the analysis method for detecting anomalies and events are employed on the processed data.

The feature selection [13] procedure usually affects the final results. Let the behaviors/features of the hosts $F = \{f_1, f_2, ..., f_M\}$ at the time $t = \{t_1, t_2, ..., t_T\}$, where N is the size of the hosts, M is the number of features and T is the length of the time series. Each host is associated with a unique identifier, which is the IP address in a computer network.

In a real computer network, the hosts are separated into the source IP (SIP) and destination IP (DIP). The traffic of the network comes from the connection among the SIP and DIP (see Fig. 1). A suspicious host (42.81.21.8) can be detected from its behaviors in the network. The connection among the hosts may vary over time in the network. For example, several users may share the same host machine or connect through the same Network Address Translation (NAT) device at some time. The hosts may change as new users to join the local network and share the same public IP, and others leave. Considering the traffic in time series, a potential event might be detected and the suspicious hosts can be traced when there is heavier traffic happened at certain moment than in normal times. We define an *event* in the network traffic as a subset of time stamps when there have high activities different from usual. The *anomalies* are the attractive hosts that have significantly changed behaviors as usual. A host is anomaly or not in the whole traffic depends on different features and time periods.

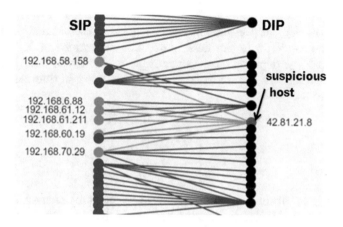

Fig. 1. The network connections and traffic details can be viewed in one single graph. The weight of the lines between the nodes indicates the traffic volume and the color of the hosts indicates the status, e.g., normal and suspicious hosts.

Anomaly detection in network traffic aims to discover a subset of the hosts that change significantly on their communication patterns. In most cases, the anomalies come with the pattern of an event happens. For example, a person visits a website at 8:00 a.m every day, it is a pattern. However, if he/she suddenly visits the same or different website many times at around 8:00 p.m. An event can be detected. Combining the uploading package amount for example, it is possible to determine the person is anomaly or not. An anomaly is usually considered as the significantly changed objects from a global aspect. However, an object may be considered as an anomaly only in a specific context but not only globally. It is found that almost all the traffic time-varying signal is multi-scales [14]. In order words, the difference between anomalous traffics and typical/normal traffics is various in different frequency bands.

3   The *SWEAT*

In the proposed method, a definition of a novelty score is given, which quantifies the similarity between two time stamps. It is used for identifying events. Eigenvectors are analyzed to spot anomalous patterns and behaviors of the hosts.

3.1   Event Detection

For the network with hosts P as IP address, the behaviors of the host at certain time t is X_{ft}^P. Define the time series in linear model for a certain feature as:

$$X_t^P = \sum_{w=1}^{\infty} a^P X_{t-w}^P + b^P \tag{1}$$

where w is the lag, or the distance between two observations, and a is the weights of the model, b is the noise part. The covariance of X_t^P and X_{t-w}^P, i.e., $cov(X_t^P, X_{t-w}^P)$ depends on the separation lag w instead of t.

Therefore, a data D in a matrix composed of N hosts in time series of length T is generated as:

$$D_{N \times T} = \begin{bmatrix} X_1^{P_1} & X_2^{P_1} & \dots & X_T^{P_1} \\ X_1^{P_2} & X_2^{P_2} & \dots & X_T^{P_2} \\ X_1^{P_3} & X_2^{P_3} & \dots & X_T^{P_3} \\ \dots & \dots & \dots & \dots \\ X_1^{P_N} & X_2^{P_N} & \dots & X_T^{P_N} \end{bmatrix} \tag{2}$$

The $D_{N \times T}$ is taken as an input (data representation) of the method, which is composed of the IP and its time series. An analysis on the data D within a moving window w is performed after then. For the time series X_t as a stationary process, $(1 \leq i \leq w, 1 \leq j \leq w)$ in the window w, the covariance C_w is obtained by:

$$C_w = \frac{1}{w} \sum_{i=1}^{w} X_i^P \cdot X_j^P \tag{3}$$

Since the X_t^P is a vector with N hosts, the covariance matrix C_w for each window is a symmetric matrix with size N. In this paper, we employ the *Pearson correlation* instead of the covariance for measuring the similarity.

$$\rho(X_i^P, X_j^P) = \frac{cov(X_i^P, X_j^P)}{\sigma_{X_i^P} \sigma_{X_j^P}} \tag{4}$$

With the moving window, there have been $T - w + 1$ lagged copies of the time series, thereby $T - w + 1$ matrices are generated.

The eigenvectors of the matrices are sorted by decreasing order of the associated eigenvalue λ_k, $0 \leq \lambda_{T-w+1} \leq \dots \leq \lambda_2 \leq \lambda_1$. As eigenvectors indicate the directions in which the principal change happens, it is natural to detect the events through the eigenvector at time t by comparing it with the eigenvectors at surrounding time stamps. A typical eigenvector $\bar{V}_t$ is calculated, which represents a typical or average behavior of all the hosts within a window w at time t. It is meant to compare the current behavior of the host at t V_t and the average behavior $\bar{V}_t$ to check if any significant change has made. A novelty score NS_t is then defined to estimate the difference by calculating the cosine distance of them:

$$NS_t = 1 - \frac{V_t \cdot \bar{V}_t}{|V_t||\bar{V}_t|} \tag{5}$$

The novelty score is a value in $[0, 1]$, where $NS = 1$ represents that there is an anomaly at that time and $NS = 0$ means the eigenvectors are the same. A high novelty score indicates an anomaly at that time point.

Input: $D_{N \times T}$, w
Output: Novelty Score NS_t
1 //Sliding a window with size w on D ;
2   FOR each $t \in (1, ..., T)$;
3   Calculate the matrices based on Eq. 4 ;
4   END FOR ;
5 Calculate the largest eigenvectors $V_{N \times (T-w+1)}$;
6 //Calculate the Novelty Scores ;
7 FOR each time t ;
8   Calculate the average eigenvector
   $\bar{V}_t = \frac{1}{w} \sum_{k=1}^{w} V_{t-k}$;
9   Novelty Score $NS_t = 1 - \frac{V_t . \bar{V}_t}{|V_t||\bar{V}_t|}$;
10 END FOR ;
11 return NS ;

Algorithm 1: Event detection in the $SWEAT$ method

The illustration of the event detection procedures is shown in Algorithm 1. Considering to put a size of w window on the input D, the expected output is the novelty score for each time stamp. As described in Fig. 2, the window is sliding on all of the hosts along time. For each time window, there is a similarity matrix composed of the values from Eq. 4. The similarity value here reflects the difference of the hosts' behavior in the time interval. The similarity matrices are multi-scale when the size w has changed and the w controls the resolution.

An event can be detected through the novelty scores NS_t when the consecutive scores, i.e., NS_{t-1}, NS_t and NS_{t+1} have a large difference than others. It is a simple way to set a threshold to detect the difference, for example, $(NS_t - NS_{t-1}) > \epsilon$ and $(NS_{t+1} - NS_t) > \epsilon$. For the window w, a set of events $E_w = \{t_1, ..., t_{topT}\}$ are thereby generated by picking the top $topT$ time stamps that have large difference. With the setting of different window size w, it is to find the events under different resolutions. The final detected events are $E = \bigcup_w E_w$.

3.2   Anomaly Detection

With the detected events in time series, it is applicable to backtrack on the anomalous hosts, who have contributed to the events. At the time that the events happen, the behavior of host p at that time E_w can be checked from the eigenvector $V_{E_w, p}$ and its related typical Eigenvector $\bar{V}_{E_w, p}$. It is to analyze if there is a linear correlation between the eigenvectors V_{E_w} and the typical Eigenvectors $\bar{V}_{E_w}$. A scatter plot can be used to see how two comparable variables, which are Eigenvector and typical Eigenvector agree with each other. A line of a so called trend-line can be drawn to be a reference, which can be fitted linearly by the line

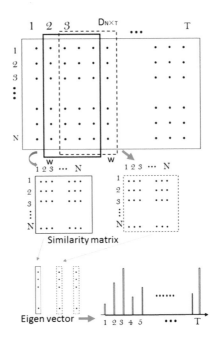

Fig. 2. The sliding window on the data and the way of calculating the similarity matrix and novelty scores.

of $y = a * x + b$. The more the two variables agree, the more the scatters tend to concentrate in the vicinity of the trend-line. If the two variables are numerically identical, the scatters fall on the trend-line exactly. Based on the scatter plot, we extract a certain number of hosts, which are the top $topIP$ farthest away from the trend-line as the anomalies, i.e., $argmax_{p=1}^{N}(dist(line, (V_{E_w,p}, V_{\bar{E_w},p})))$, where p is the index of the IP addresses. The anomalies detected at different window w are aggregated into one suspicious IP list.

For the proposed method, the time complexity mainly comes from the calculation of the similarity matrix and its eigenvector. The similarity matrix step takes $O(N^2 \times w)$, where N is the node size and w is the window size. To obtain its eigenvector, the fastest implementation takes $O(N^2)$. It totally takes $O(N^2 * T * w * s)$ for the event detection step, where the s represents the scale levels. Therefore, the time and space complexity for the method is $O(N^2)$.

4   Evaluation on the *SWEAT*

In order to validate and evaluate the proposed method, we perform experiments firstly by testing the method on a network traffic data. The tested data is from a network security technology company, which is collected from June 15, 2015 to June 28, 2015. The raw data is preprocessed and feature engineering has been performed on the data for further feature selection. The TCP flow part of the raw data is selected as the exact test data in this paper. After a rough selection, four features from the DIP, which are the amounts of upload package, download package, FIN and RST in one hour, are extracted from the raw data for further analysis. Since the traffic of the network happens at anytime between any IP addresses, there have been a lot of zeros in the original data. Hence, we have two ways to pre-process the original data and generate two kinds of testing data TCP_1 with 336 time stamps, which contains up_ts1_60, $down\_ts1\_60$, fin_ts1_60, rst_ts1_60, $up\_down\_ts1\_60$, and TCP_2 with 84 time stamps, which contains up_ts4, $down\_ts4$, fin_ts4, rst_ts4, up_down_ts4. The data in TCP_1 is obtained by filtering out the IPs with certain proportion of zeros in one hour. The data in TCP_2 is extracting the features of the traffic within four hours. The $up\_down\_ts1\_60$ and up_down_ts4 are generated from the total amount of upload and download package.

Because of security issues, there is not an exact blacklist for the whole network, which is quite huge. However, the labels (anomaly or not) of the IP addresses can be verified manually from the database of the company. Hence, limited evaluation can be done on the accuracy of the proposed method. Totally, there are 1583 IP addresses in the testing data, among which, 202 IP addresses are anomalies.

4.1 Detected Events and Suspicious IP Addresses

One of the goals of the proposed method is to detect the events in the data. The events are the moments that significant changes have been made at different scale level. The sliding window size is changed from one to 24. As shown in Fig. 3, the novelty scores obtained from different window sizes on data up_down_ts4 of the total upload and download package amount are demonstrated. According to the distribution of the novelty scores, it reveals how the IPs behave at different resolutions. The higher the novelty scores are, the high possibility exists anomaly. The events correspond to local maximas of the novelty scores at different window sizes, which are identified by simple analysis of each score's second derivatives. For example, as shown in the Fig. 3, the largest local maxima is at the time 50 when the window size is 21. It indicates that the window sizes are helpful in finding events at different resolutions.

The histogram for the data up_down_ts1 with the total upload and download package amount is shown in Fig. 5, where six events are labeled. According to the histogram, the events mostly happen in the morning (7:00–8:00) or evening (18:00 and 23:00). Comparing the Figs. 4 and 5, the result (more events) from the data up_down_ts1 obtains more details as the resolution of the data is higher.

An event in the network traffic should be produced by the hosts. After the events being detected, the hosts that are related to the events can be found. A scatter plot is used to analyze the Eigenvector and typical Eigenvector at the event time. As shown in Fig. 6, scatter plots on the Eigenvector and typical Eigenvector at the event for up_down_ts1 and up_down_ts4 direct to the suspicious hosts. A fitted line for the Eigenvector and typical Eigenvector is plotted. Since the typical Eigenvector reflects the average behavior of the hosts in a time

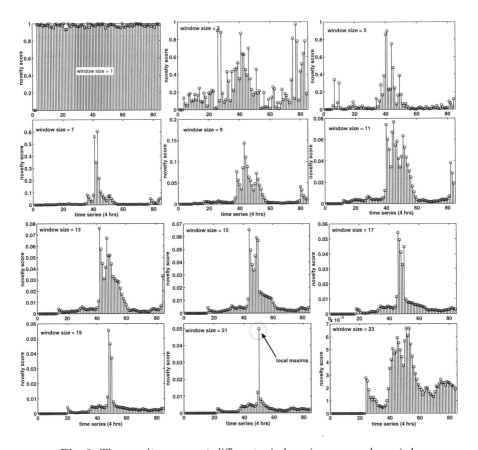

Fig. 3. The novelty scores at different window sizes on *up_down_ts4*.

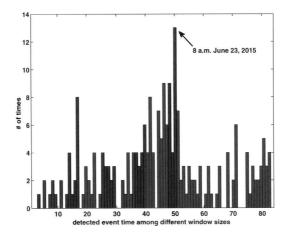

Fig. 4. A histogram of the detected top one event time stamps among different window sizes on *up_down_ts4*.

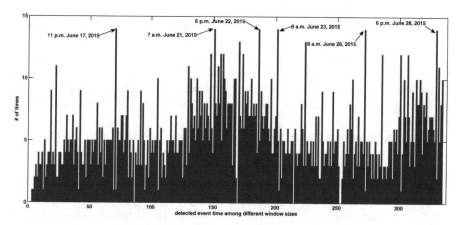

Fig. 5. A histogram of the detected top six event time stamps among different window sizes on *up_down_ts*1.

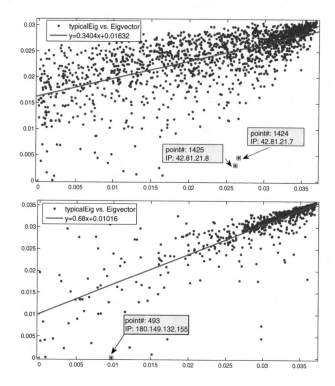

Fig. 6. Scatter plots drawn on two variables: the Eigenvector (*Eigvector*) and typical Eigenvector (*typicalEig*) at the event (time stamp 50) for *up_down_ts*4 (up) and those at the time stamp 71 for *up_down_ts*1 (down).

period, the scatter plot of the two variables shows how the hosts are different from their typical behaviors. Those points far away from the fitted line are considered the anomalies, who have contributed to the event. The two points no. 1424 and no. 1425 with the IP address 42.81.21.7 and 42.81.21.8 respectively are suspicious IP addresses from the first scatter plot. Meanwhile, the point no. 493 with the IP 180.149.132.155 is considered as the suspicious IP from the second scatter plot.

There have been two parameters in the proposed method, which is the number of events $topT$ and the number of anomalies $topIP$. It is interesting to know that how the selection of $topT$ and $topIP$ affects the accuracy of the method. Taking the data up_down_ts4 and a certain size of sliding window, an experiment is conducted on different settings of $topT$ and $topIP$. According to the result, the effect of the $topT$ on accuracy is quite little. However, the selection of $topIP$ affects much on the accuracy. For the anomaly detection problem in network traffic, selecting independent, informative and relevant features improves the detection quality for the algorithms. In our case, we study on the four features of the TCP flow. An experiment on how features affect the accuracy is also conducted. The numbers of correctly detected IP addresses from different features vary, which indicates that the selection of features affects the accuracy of the detection algorithm. According to the numbers, the features in one hour bring better result than those in four hours in general. Because of security issues, there is not an exact blacklist for the whole network, which is quite huge. In order to validate and evaluate the method, the AMPLIFIER, which can be visited from idatatongji.com/demos/amplifier/, a visualization tool is designed based on the analysis method. It can demonstrate the analysis procedures, detect security incidents and anomalous hosts in time series of the network traffic. Special glyphs are designed for the network traffic data in the tool. Interactions can be performed on the tool for further visualizing the behaviors and patterns of the hosts.

5   Conclusion

The analysis of the anomalies in a network traffic is a challenging problem. In this paper, we introduce a sliding window based anomaly detection method for the time series of network traffic. The method includes a multi-scale detection through the setting of sliding window sizes. It can detect the events in the network traffic, meanwhile, the anomalous hosts who contribute to the event can also be detected. The method has been validated and evaluated through experiments.

Acknowledgement. This work is partially supported by National Natural Science Foundation of China (Grant No. 61702372, No. 61672380) and The Fundamental Research Funds for the Central Universities of China.

References

1. Huang, H., Al-Azzawi, H., Brani, H.: Network traffic anomaly detection. Eprint Arxiv (2014)
2. Gupta, M., Gao, J., Aggarwal, C., Han, J.: Outlier detection for temporal data: a survey. IEEE Trans. Knowl. Data Eng. **26**(9), 2250–2267 (2014)
3. Ide, T., Kashima, H.: Eigenspace-based anomaly detection in computer systems. In: Proceedings of the 10th ACM SIGKDD, pp. 440–449, August 2004
4. Akoglu, L., Faloutsos, C.: Event detection in time series of mobile communication graphs. In: Proceedings of Army Science Conference, vol. 2 (2008)
5. Ahmed, M., Mahmood, A., Hu, J.: A survey of network anomaly detection techniques. J. Netw. Comput. Appl. **60**, 19–31 (2015)
6. Monowar, H., Bhuyan, D., Bhattacharyya, K., Kalita, J.: Network traffic anomaly detection: methods, systems and tools. IEEE Commun. Surv. Tutor. **16**(1), 303–336 (2014)
7. Lakhina, A., Crovella, M., Diot, C.: Diagnosing network-wide traffic anomalies. ACM Sigcomm Comput. Commun. Rev. **34**(4), 219–230 (2004)
8. Novakov, S., Lung, C., Lambadaris, I., Seddigh, N.: Combining statistical and spectral analysis techniques in network traffic anomaly detection. In: 2012 Next Generation Networks and Services, pp. 94–101 (2012)
9. Kaur, G., Saxena, V., Gupta, J.: A novel multi scale approach for detecting high bandwidth aggregates in network traffic. Int. J. Secur. Appl. **7**(5), 81–100 (2013)
10. Kwon, D.W., Ko, K., Vannucci, M., Reddy, A.L.N., Kim, S.: Wavelet methods for the detection of anomalies and their application to network traffic analysis. Qual. Reliab. Eng. Int. **22**(8), 953–969 (2006)
11. Gao, J., Fan, W., Turaga, D., Verscheure, O.: Consensus extraction from heterogeneous detectors to improve performance over network traffic anomaly detection. In: Proceedings - IEEE INFOCOM, vol. 267, no. 2, pp. 181–185 (2011)
12. Mansman, F., Meier, L., Keim, D.A.: Visualization of host behavior for network security. In: VizSEC 2007, pp. 187–202 (2007)
13. Iglesias, F., Zseby, T.: Analysis of network traffic features for anomaly detection. Mach. Learn. **101**, 59–84 (2015)
14. Br, B.: Multi-scale analysis and modeling using wavelets. J. Chemom. **13**(3–4), 415–434 (1999)

The Grid-Based Spatial ARIMA Model: An Innovation for Short-Term Predictions of Ocean Current Patterns with Big HF Radar Data

Ratchanont Pongto[1], Nopparat Wiwattanaphon[1], Peerapon Lekpong[1],
Siam Lawawirojwong[2], Siwapon Srisonphan[3], Kerk F. Kee[4],
and Kulsawasd Jitkajornwanich[1(✉)]

[1] Department of Computer Science, Faculty of Science,
King Mongkut's Institute of Technology Ladkrabang, Bangkok 10520, Thailand
{57050313,57050254,57050297,kulsawasd.ji}@kmitl.ac.th
[2] Geo-Informatics and Space Technology Development Agency (PO),
Bangkok 10210, Thailand
siam@gistda.or.th
[3] Department of Electrical Engineering, Faculty of Engineering,
Kasetsart University, Bangkok 10900, Thailand
fengspsr@ku.ac.th
[4] College of Media and Communication, Texas Tech University, Lubbock,
TX 79409, USA
kerk.kee@gmail.com

Abstract. Marine natural disasters have direct impacts on countries as well as their residents living on and near the coast. Warning and monitoring system can aid in reducing the loss of lives in the event of a disaster. HF (high frequency) radar, an IoT-enabled ocean surface current monitoring system, implementation is one of the first attempts towards achieving this goal. Although HF systems can monitor sea current patterns in terms of speed and direction for each of the pixels of the coverage area, it fails to predict future values, which are essential to many applications such as oil-spill trajectory prediction (using the GNOME suite: General NOAA Operational Modeling Environment), water quality control and management, and optimized sea navigation. In this paper, we propose a model, called the grid-based spatial ARIMA (auto-regressive integrated moving average), to estimate the forecast values. As a result, the full potential of the HF systems can be utilized. The method considers not only observations of POI (point of interest), but also its neighboring pixels when predicting future values. The proposed method is implemented and compared with other existing approaches, including baseline, kNN, traditional ARIMA model, and LSTM (long short-term memory) techniques. The experimental results showed that our approach outperformed other methods in V comp prediction (with RMSEs of 6.23265) with a configuration of $(2, 0, 1)$ as (p, d, q) and a historical dataset of 1 day and 7 h prior. This configuration was found to be the best combination.

Keywords: Ocean surface current · ARIMA · HF radar ·
Spatio-temporal data mining · GNOME · Big data

© Springer Nature Switzerland AG 2020
P. Boonyopakorn et al. (Eds.): IC2IT 2019, AISC 936, pp. 26–36, 2020.
https://doi.org/10.1007/978-3-030-19861-9_3

1 Introduction

Marine natural disasters have direct impacts on countries as well as their residents living on and near the coast. Warning and monitoring system can aid in reducing the loss of lives in the event of a disaster. HF (high frequency) radar station implementation is one of the first attempts towards achieving this goal. For example, 18 HF coastal radar stations were installed along the Gulf of Thailand in order to monitor marine natural disasters as well as relevant risks by retrieving various parameters of sea surface, such as wave data and height, surface current data and direction, and current acceleration, covering the sea body 10–60 km away from the coast [1]. The HF systems record ocean surface current information on an hourly basis, including the volume of current body mass and the direction of the current.

Although HF systems can monitor sea surface patterns in terms of speed and direction for each pixel of the coverage area, it fails to predict future values, which are essential to many applications such as oil-spill trajectory prediction [2] (using GNOME), water quality control/management, and sea navigation optimization. We propose a model, called grid-based spatial ARIMA (auto-regressive integrated moving average), to estimate the forecast values so that the full potential of HF systems can be utilized. The traditional ARIMA is modified to better suit our data configuration with which spatial and temporal dimensions of data are incorporated. The data sets are stored and processed by using a cluster of servers, Windows Server 2012, MS SQLServer, Apache2 and python scripts, and the results are visualized by using JSON, Node.js, Google Maps APIs as well as HTML, CSS, Javascript and PHP.

The organization of this paper is as follows. Section 2 discusses related work in spatio-temporal mining and prediction. Section 3 describes HF radar data properties and characteristics, and it demonstrates the proposed methodology followed by Sect. 4, which presents the experimental results. Finally, we conclude our work and discuss some future work in Sect. 5.

2 Related Work

Cheng et al. [3] surveyed several applications of spatio-temporal data mining and knowledge discovering (STDMKD) with an emphasis on wildfire forecasting. In their research, a wildfire forecasting system is implemented based on wildfire events occurred in a given area in Canada using three different methods, namely, ISTIFF, ARIMA, and STIFF, which are then evaluated and compared against each other. The result shows that ISTIFF has the highest accuracy.

According to [4], to accommodate large-scale spatio-temporal data analysis, the concepts of spatio-temporal data mining and related tools need to be fully understood. A long list of applications can be benefited from spatio-temporal data mining technique such as ecology system and environmental management security, transportation, geography, spread of disease, and climate and weather. The main takeaway from this work is that domain knowledge plays a very important role in tuning and perfecting algorithms for maximum efficiency.

Longitude	Latitude	Ucomp	Vcomp	VectorFlag	UStdDev	VStdDev	Covariance
100.1022264	12.2539686	10.533	-6.536	0	4.95	3.86	-17.18
100.1206079	12.253999	-1.258	-14.391	0	3.47	2.65	-7.76
100.1389894	12.2540281	-5.002	-15.802	0	3.32	2.92	-8.32
100.1573709	12.254056	-7.461	-15.855	0	4.96	4.9	-23.1
100.1757525	12.2540825	-8.445	-15.344	0	6.06	6.56	-38.59
100.194134	12.2541078	-6.314	-12.525	0	5.35	6.51	-33.31
100.230897	12.2541544	0.148	-4.985	0	5	6.24	-28.66
100.2492786	12.2541758	-1.812	-9.258	0	5.26	7.99	-39.39
100.2676601	12.2541959	-2.031	-7.061	0	4.14	6.13	-23.25
100.2860416	12.2542147	-1.534	-3.541	0	2.55	2.62	-5.4
100.3044232	12.2542322	-4.322	-4.143	0	999	999	999

Fig. 1. An example of TUV file and corresponding reported grids from HF radar

In [5], not only were spatio-temporal data models proposed, indexing and querying methods were also developed. Indexing and querying help improve data retrieval performance as well as mining process efficiency as a whole. Spatio-temporal concepts are quite different from concepts that are only spatial or temporal in nature. For example, the spatial indexing focuses on static object while temporal indexing focuses on time series without the object shape bounding. Thus, the indexing for spatio-temporal data (e.g., data of moving objects) need to incorporate both the dimensions of time and space. There are two main approaches presented in this research; both are the variation of R-tree and grid-based indexing.

Spatio-temporal data involve describing the position and the status of objects at different time intervals. Spatio-temporal data can be found in many applications, such as traffic monitoring, environmental management, and weather forecasting. These applications collect data at different places and times in various formats, which in turns lead to challenges in processing and analyzing "big spatio-temporal data". Some examples of large-scale spatio-temporal data include GPS information of a moving car, movement of a hurricane, and historical patterns of coastal changes. The spatio-temporal data can capture changes of spatial features (e.g., shape) at location x over time interval t.

3 Methodology

In this section, we describe our coastal radar data properties, followed by our grid-based spatial ARIMA model according to traditional ARIMA time series analysis.

3.1 Coastal Radar Data Description

The two main characteristics of ocean surface current are direction and velocity, which were used in the analysis in this paper. The data set was provided by Coastal Radar Data Center at Geo-Informatics & Space Technology Development Agency (Public Organization), Ministry of Science & Technology of Thailand (http://coastalradar. gistda.or.th). The data set was collected from January 2014 to October 2015. It is in TUV format, where the spatial information is embedded in the text file. The data were reported hourly and stored in a fixed grid format. There is a total of 15,799 files (hours) with 3,081 grid locations (lat/lon) observed in our data set. Each measurement (row) represents an average speed and a direction of the sea surface current over a 2×2 km boundary centered at the reported grid (pixel). A TUV file contains 21 columns based on WGS84 reference system and UTC + 0 (Atlantic/Reykjavik) time zone. The example of TUV file and corresponding study area are shown in Fig. 1. As

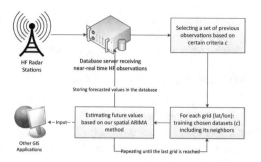

Fig. 2. An overview of our methodology

shown in the TUV file, many variables are captured by the HF radar system. Out of 21 attributes, only six of them are used in this research: Longitude in degrees (LOND), Latitude in degrees (LATD), Ucomp (VELU: current vector velocity eastern (X-axis) component in cm/s), Vcomp (VELV: current vector velocity northern (Y-axis) component in cm/s), overall Velocity (VELO) and Direction (HEAD) [8]. The time attribute is also used, which was indicated in the filename. These seven measures are selected because they represent direction and velocity, the two main ocean current observation characteristics that are needed for other significant GIS applications, including oil-spill trajectory tracking (using GNOME), search and rescue, and optimized sea navigation or even hydrological disaster prediction.

3.2 ARIMA Time Series Analysis

ARIMA (auto-regressive integrated moving average) method, developed by George E. P. Box and Gwilym M. Jenkins in 1970, was used to analyze time series data for this paper. It is considered one of the most popular tools for short-term prediction; its RMSE rate is typically lower than other methods (such as trend analysis, exponential smoothing, and multiple regressions). Also, it is less complicated in modeling predictive system as multi-tier equations. Because ARIMA estimate weights for predicting future values based on previous time-point data, it is important to eliminate trends and seasonality existed in the data, which can be done in various ways, such as taking log derivatives and calculating means. The data are required to be "stationary," a non-biased form that is ready for analyzing the movement of the data at a desired point in space and surrounding areas. The ARIMA model consists of three main parts: (1) auto regressive model (AR), (2) moving average model (MA), and (3) integrated process (I).

Auto Regressive Model (AR). Auto-regressive model describes how y_t observation value can be determined by $y_{t-1}, \ldots, y_{t-p}$ (or previous p observation values). The AR (p) process can be written in an equation with rank p as follows [6]:

$$AP(p) : x_t = \mu + \emptyset_1 x_{t-1} + \emptyset_2 x_{t-2} + \cdots + \emptyset_p x_{t-p} + \varepsilon_t \qquad (1)$$

Moving Average Model (MA). The moving average (MA) model estimates observed y_t value by focusing on the observation different or error values from ε_{t-1}, …, ε_{t-q} (or preceding q tolerance). The MA(q) or moving system can be written in term of q [6]:

$$MA(q) : x_t = \mu + \varepsilon_t - \theta_1 \varepsilon_{t-1} - \theta_2 \varepsilon_{t-2} - \ldots - \theta_q \varepsilon_{t-q} \tag{2}$$

Integrated Process (I). The integrated process depicts a time difference between current data and historical data in the last d periods. This process is necessary when the given time series are not stationary as ARIMA model only supports stationary dataset. In case of non-stationary data, it must be converted to stationary before ARIMA model can be applied. The process I(d) can be described as follows [6]:

$$\Delta_d x_t = \Delta_{d-1}(x_t - x_{t-1}) \tag{3}$$

Auto-regressive moving average (ARMA) model combines auto regressive (Eq. 1) and moving average processes (Eq. 2) together. It considers auto regressive system with p observations with moving average of q, with an assumption of data being stationary. The ARMA model can be expressed as an equation as follows:

$$y_t = \delta + \emptyset y_{t-2} + \ldots + \emptyset y_{t-p} + \varepsilon_t - \theta_1 \varepsilon_{t-1} \tag{4}$$

ARIMA (auto-regressive integrated moving average model) instead, consists of three processes mentioned above (Eqs. 1–3), which can be defined in terms of (p, d, q) as:

$$\Delta_d y_t = \delta + \emptyset \Delta_d y_{t-1} + \emptyset \Delta_d y_{t-2} + \ldots + \emptyset \Delta_d y_{t-p} + \varepsilon_t - \ldots - \theta_q \varepsilon_{t-q} \tag{5}$$

Our spatial ARIMA model is based on the ARIMA model as opposed to the ARMA model. While the traditional ARIMA uses a single POI (lat/lon) when calculating future values, the proposed grid-based spatial ARIMA model utilizes the neighboring pixels as well as the coordinate point of interest, so the reliability of results can be enhanced. This is based on the assumption that a pixel and its surrounding ones should have similar movement patterns. We believe that this approach is a methodological innovation. An overview of our method is shown in Fig. 2.

Additionally, according to [7], their ARIMA analysis is solely determined by the previous t time points when predicting $t + 1$ value. However, considering only the previous values may not be sufficient for capturing the real and actual patterns of moving currents in the ocean. Thus, in our method, we chose various ranges of data covering before and after the time point of interest for the experiments. The dataset is stored in a Microsoft SQL Server database with spatial features for easy neighboring pixels retrieval. The conceptual design of our database is shown in Fig. 3.

4   Experimental Design and Results

In our experiments, we estimated and compared the results with other existing methods (baseline, kNN, traditional ARIMA, and LSTM—with ~1,021 sites) using the same selected datasets. Each of the predicted values (i.e., U comp, V comp, velocity, direction) of these four data points is evaluated using RMSE and accuracy.

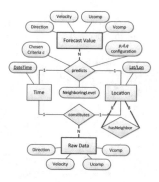

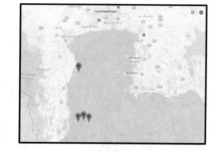

Fig. 3. Our conceptual storage structure   **Fig. 4.** The selected four data points

Five main factors are considered: p, d, q, neighboring pixel level, as well as a chosen range of historical data (i.e., the last 7, 15, 24 h within the last 1, 7, 10, and 15 days). To fully represent the complex characteristics of the complete data set, a small subset of data of four sites may not be realistic enough for modeling complex spatio-temporal phenomena. Therefore, we performed the experiments with a greater range of datasets (~ 350 sites) in order to capture the complexity under the actual environment and to maximize the real performance and accuracy of our predictive model.

Tables 1, 2, 3 and 4 are amongst other experiments we performed, describing how the "best" predicted U comp, V comp, velocity and direction values of the four data points were selected, after the experiments using all the possible combinations of different ranges of historical datasets (time periods and number of days). The actual recorded values were listed in the leftmost column for assessing accuracy of predictions. We selected the best prediction of the four values based on the lowest RMSEs among the ranges (in red boundaries). From column two onwards, different configurations of (p, d, q) are tested with the level of neighboring pixel = 1 and the range that provided the best prediction is listed in the table caption. In the same fashion, Tables 5, 6, 7 and 8, the RMSEs of Tables 1, 2, 3 and 4, were calculated. The four data points consisted of following (lat/lon with SRID = 4326): (12.9229064, 100.1194929), (12.253999, 100.1206079), (12.2540281, 100.1389894), and (12.254056, 100.1573709).

The map visualization of the data points is shown in Fig. 4. For ease and convenience in interpreting results of Tables 1, 2, 3 and 4, we color-coded the results using an accuracy scale, with green being the most accurate and red being the least accurate while gray representing the undetermined. Similarly and in Tables 5, 6, 7 and 8, we color-coded the numerical values of the predicted results using a validity scheme, where the most valid predicted results are based on obtaining values from all four sites shown in Fig. 4. In other words, black = validity achieved with 4 data points, green = 3 data points, purple = 2 data points, red = 1 data point, and no color = 0 data point.

In Table 9, all the performed experiments were combined and compared to with previous approaches [7] using the greater range of datasets (~ 350 sites). Figure 5 shows the screenshot of the actual implementation of our predictive system for the Gulf of Thailand, visualized on the map, including both direction (arrows) and velocity

Table 1. Best U comp prediction with the last 7 h within the last 10 days of data

Real	1.0.1	1.0.2	1.0.3	1.0.4	1.0.5	1.1.1	1.1.2	1.1.3	1.1.4	1.1.5
0.832	0.6365	0.5348				-0.0061				
4.252	2.2294					2.5266	2.7775	1.3284		
-4.428	1.5501	1.593			0.7328	1.5193	1.4773	1.4822	-0.6842	-0.0248
-0.205	1.8492			0.4829		1.6645	1.6963	1.2574	0.3034	0.2996
Real	2.0.1	2.0.2	2.0.3	2.0.4	2.0.5	2.1.1	2.1.2	2.1.3	2.1.4	2.1.5
0.832	0.7487	1.0880			0.9081	0.0182				
4.252	2.1349	4.1391				2.4508	2.7449			
-4.428	1.6826	0.3602				1.4230	-0.0481	-1.9748		0.3627
-0.205	0.4870	-1.8336		-0.1929		1.4196	0.1370			
Real	3.0.1	3.0.2	3.0.3	3.0.4	3.0.5	3.1.1	3.1.2	3.1.3	3.1.4	3.1.5
0.832	0.8925	0.5232	0.0807	0.2821		0.0361	0.3373			
4.252	3.1423					7.9026				
-4.428	1.8263	2.2530	-1.1654		-0.7320	1.3782	-0.3188	2.3374	-2.4852	-1.7593
-0.205	-0.8435	0.2824	-0.8668	-0.3805		0.6069	0.7828			
Real	4.0.1	4.0.2	4.0.3	4.0.4	4.0.5	4.1.1	4.1.2	4.1.3	4.1.4	4.1.5
0.832	0.9977	0.8627	0.1040	1.8935		0.1658				
4.252	3.4451	6.2698	2.8311			8.3083				
-4.428	0.0354	-2.0829	-1.1121	2.3181	2.5157	0.0480	-1.6341	-2.2695		-1.7531
-0.205	-2.3049	-0.2023	-0.9412	-0.8398	-0.8921	-1.2548				
Real	5.0.1	5.0.2	5.0.3	5.0.4	5.0.5	5.1.1	5.1.2	5.1.3	5.1.4	5.1.5
0.832				1.1219	2.3957	0.1213		-0.2391		
4.252	4.0275	0.8544	5.1259	0.0352	0.2540	4.6881	5.4211			
-4.428	0.1846		-0.3571		2.0129	0.0632	-1.6154	1.8766		-2.7999
-0.205	-1.3676	-0.9282	-1.0101	-1.6360	-0.8016	-1.2725				

Table 2. Best V comp prediction, with the last 7 h within the last 7 days of data

Real	1.0.1	1.0.2	1.0.3	1.0.4	1.0.5	1.1.1	1.1.2	1.1.3	1.1.4	1.1.5
-21.466	-1.344	8.115	-23.100		-27.839	-4.782				
-28.081	-6.003	-4.015				-0.391	-9.448		-23.726	-30.949
-32.392	-7.273		-27.265	-22.555	-24.483	-8.115	-10.285	-30.507	-44.311	-34.773
-26.654	-7.015	-3.848	-26.867	-41.271	-36.352	-9.906	-10.181	-30.607	-38.861	-31.884
Real	2.0.1	2.0.2	2.0.3	2.0.4	2.0.5	2.1.1	2.1.2	2.1.3	2.1.4	2.1.5
-21.466	-14.055		-15.204	-2.029	-27.881	-17.034				
-28.081	-6.198	-7.233	4.331			-22.045	1.465	-12.918		
-32.392	-16.628			-16.346	833.070	-18.621	-30.428		-36.660	
-26.654	-15.651	17.582		-18.347	-27.055	-17.334	-31.084	-28.691	-39.041	-35.210
Real	3.0.1	3.0.2	3.0.3	3.0.4	3.0.5	3.1.1	3.1.2	3.1.3	3.1.4	3.1.5
-21.466	-17.158		-2.099	-33.875	-33.323	-35.133	-34.896	-20.461		
-28.081	-5.380	-4.917	-3.272			14.395	1.173	-3.418	95.949	
-32.392	-23.543	-23.302		-13.459		-42.090	38.508	655.608		
-26.654	-25.450	-62.370		-14.671		-43.195		-46.603		
Real	4.0.1	4.0.2	4.0.3	4.0.4	4.0.5	4.1.1	4.1.2	4.1.3	4.1.4	4.1.5
-21.466	-11.733	-4.202	-0.292			-16.120	20.549			
-28.081	-1.790	-4.369	0.981			-14.288	-1.323		-22.540	
-32.392	-23.538	11.670		-12.785	-19.684	-37.844	-42.119		2592.032	
-26.654	-25.534	-12.646		-14.027			-42.700	-32.227		
Real	5.0.1	5.0.2	5.0.3	5.0.4	5.0.5	5.1.1	5.1.2	5.1.3	5.1.4	5.1.5
-21.466	-22.674	-4.251	-0.002	-21.097						
-28.081	3.995	-10.467	-0.414			30.918				
-32.392		-26.787	-16.847		32.786					
-26.654	-129.272	-140.173		-16.783	-27.918					

Table 3. Best direction prediction, with the last 7 h within the last 7 days of data

Real	1.0.1	1.0.2	1.0.3	1.0.4	1.0.5	1.1.1	1.1.2	1.1.3	1.1.4	1.1.5
177.8									188.97	
171.4	164.52	159.70				154.35	161.84	157.51	173.14	
187.8	159.48					150.55	156.79	157.14		
180.4	159.68					144.42	163.56	161.41		163.94
Real	2.0.1	2.0.2	2.0.3	2.0.4	2.0.5	2.1.1	2.1.2	2.1.3	2.1.4	2.1.5
177.8	157.42						309.53			191.78
171.4	156.80	153.15				155.54	161.30	121.54	122.60	
187.8	178.15	177.25				155.20	155.69	159.73		
180.4	200.91	178.41				155.79	159.13			
Real	3.0.1	3.0.2	3.0.3	3.0.4	3.0.5	3.1.1	3.1.2	3.1.3	3.1.4	3.1.5
177.8						210.19	183.74	178.30	176.78	195.93
171.4						111.67				
187.8	178.18	181.38				153.52			148.55	
180.4	199.25	209.42				162.65	162.73			
Real	4.0.1	4.0.2	4.0.3	4.0.4	4.0.5	4.1.1	4.1.2	4.1.3	4.1.4	4.1.5
177.8	171.76	202.58	694.94			211.07	194.47	174.75	207.06	
171.4	166.95		160.80			148.98	159.92	147.77		
187.8	172.64	178.58				156.41				
180.4	188.96					168.25	92.78			
Real	5.0.1	5.0.2	5.0.3	5.0.4	5.0.5	5.1.1	5.1.2	5.1.3	5.1.4	5.1.5
177.8				200.44		197.06	202.22	202.69	200.18	
171.4	160.84	164.73				142.89	159.75	134.55		
187.8	174.20	154.00	184.74		463.08	158.52	121.55	126.80		
180.4	193.37	166.57	197.01	172.55	184.69	173.35	173.60			

Table 4. Best velocity prediction, with the last 7 h within the last 15 days of data

Real	1.0.1	1.0.2	1.0.3	1.0.4	1.0.5	1.1.1	1.1.2	1.1.3	1.1.4	1.1.5
21.482	7.870	-11.631	19.372	19.828	18.487	9.979		10.767		
28.401	12.814	12.707					24.365	22.671		
32.693	11.430	16.257	15.570	22.520	20.200	12.994			21.313	21.160
26.655	12.739	17.488	17.801	26.142	22.697	13.320		26.764	24.481	23.774
Real	2.0.1	2.0.2	2.0.3	2.0.4	2.0.5	2.1.1	2.1.2	2.1.3	2.1.4	2.1.5
21.482	7.088	12.768	19.989	18.694	18.421	21.326	16.431	16.722		18.731
28.401	12.675	12.725	16.041			18.581	21.792	21.774		
32.693	13.254	15.844	22.790	18.777	18.758	17.602	18.622	19.482	21.142	23.329
26.655	13.257	17.694	25.090	22.162	21.454	18.852	20.440	21.350	19.390	25.545
Real	3.0.1	3.0.2	3.0.3	3.0.4	3.0.5	3.1.1	3.1.2	3.1.3	3.1.4	3.1.5
21.482	11.876	16.161			18.541	11.204	16.422	16.893	23.189	18.485
28.401	16.401		20.833			20.081	20.549	21.794		19.478
32.693	17.389	18.394	19.412	19.576	22.723	17.611	18.666	20.071	22.073	22.771
26.655	18.618	19.836	20.775	19.156	24.962	18.583	20.241	21.607	23.676	25.554
Real	4.0.1	4.0.2	4.0.3	4.0.4	4.0.5	4.1.1	4.1.2	4.1.3	4.1.4	4.1.5
21.482	11.318	16.145	16.778			18.237	15.954	18.135	18.503	17.971	18.240
28.401	19.216	19.659	20.825	22.191		26.634	21.179	20.970		
32.693	17.382	18.279	19.606	22.216	21.650	22.385	24.099	25.720	23.818	23.455
26.655	18.348	19.681	20.993	23.338	24.725	24.512	26.772	28.215	25.477	25.541
Real	5.0.1	5.0.2	5.0.3	5.0.4	5.0.5	5.1.1	5.1.2	5.1.3	5.1.4	5.1.5
21.482	16.699	17.058	18.378	17.434	18.111	16.807	18.976	17.359	18.670	
28.401	19.068	20.187	21.461	22.200		20.479	20.855		21.728	
32.693	22.190	23.736	25.216	23.302	19.646	24.026	25.635	23.779	20.945	22.357
26.655	24.213	26.373	27.704	24.838	24.900	26.134	27.967	28.113	28.566	

(colors). In Tables 1, 2, 3 and 4, the "Real" column describes the true recorded values of the date of 2014-04-01 04:00 from the four coordinates given above, and the other columns are values predicted by our grid-based spatial ARIMA method. The tables show predicted results, with the cell's color-coding schemes as explained previously. Tables 5, 6, 7 and 8 show the RMSE values from the predicted results. Each column shows the RMSE value derived from the predicted results compared to the actual values. The colors of the cell's background and the numerical values make it easier for readers to assess how close the predicted results are to the actual recorded values.

Table 5. Corresponding RMSEs of Table 1

n	_,0,1	_,0,2	_,0,3	_,0,4	_,0,5	_,1,1	_,1,2	_,1,3	_,1,4	_,1,5
1	3,319928	4,262743		0.68797	4,750834	3,261413	3,681566	3,89942	2,662842	3,133893
2	3,248184	2,532684		0.012067	0.076134	3.192964	2.681543	2.453163		4.790708
3	3.085595	2.722815		0.470844	3.67506	3.373647	2.45403	2.96390	1.942791	2.668733
4	2.500514	1.547557	1.87656	4.95079	4.933006	3.470494	2.793901	2.158508		2.67486
5	2.789447	2.456289	2.172278	2.576363	3.881602	2.34549	2.15375	4.521926		1.634062

Table 6. Corresponding RMSEs of Table 2

n	_,0,1	_,0,2	_,0,3	_,0,4	_,0,5	_,1,1	_,1,2	_,1,3	_,1,4	_,1,5
1	21.58921			12.53823		21.72462	69.2118	3.094488		
2	15.03155	34.57924	23.3423	15.32558		9.113215	15.61935	16.8178	9.76613	8.556198
3	12.47487		22.25524	14.80161	11.85689	13.12106	18.36385	344.3659	124.0594	
4	14.43916	19.12125	25.42574	16.49067	12.70835	9.102101	16.34721	5.573263	1856.177	
5		58.13133		6.984873	13.89268	17.16301				

Table 7. Corresponding RMSEs of Table 3

n	_,0,1	_,0,2	_,0,3	_,0,4	_,0,5	_,1,1	_,1,2	_,1,3	_,1,4	_,1,5
1	20.69558	1.70015					20.88961	22.31387	130.3273	16.46071
2	16.75983					25.30119	25.26904	40.45762	48.83028	13.97545
3	14.95325	21.014494				39.07298	13.18333	27.75728	1.096716	18.12848
4	9.07044	18.69534	365.7504			26.17372		109.2959	29.26125	
5		21.43314	11.94654	16.94416	460.2542	22.86124	35.94196		22.37872	

Table 8. Corresponding RMSEs of Table 4

n	_,0,1	_,0,2	_,0,3	_,0,4	_,0,5	_,1,1	_,1,2	_,1,3	_,1,4	_,1,5
1	16.37965	13.20535	13.07938		7.728154			8.611038	7.461217	8.405572
2	15.7065	13.09519	7.992525	8.595086	8.767497	11.36679	9.467817	8.040781	8.758271	6.860348
3	11.56589	9.74759	9.455931			10.81063	9.016102	7.897835		
4	11.03461	9.504031	8.409419	7.28770		7.104521	5.857793	5.365601		
5	7.473543	6.32139	5.366525	6.048311	7.845843	6.325106	5.366067		7.349952	10.33634

Note: each cell id is n,m,l where n is the row number, $m=0$ for the first five columns and $m=1$ for the last five, and l is the column index shown in the header.

When we compared the results from our grid-based spatial ARIMA model with other methods, including baseline, KNN, traditional ARIMA, LSTM, the results showed that our approach outperformed other methods for V component prediction. In addition, we found that selecting the right range of data set had a direct effect on the accuracy of the prediction (see the red rectangular borders in Tables 1, 2, 3, 4, 5, 6, 7 and 8). Particularly, for U comp prediction, a combination of (p, d, q) as $(4, 0, 2)$ respectively, using 10 days of dataset with 7 h prior produced the best prediction of RMSE = 1.547556, while for V comp prediction, a combination of (p, d, q) as $(2, 0, 1)$, using 7 days of dataset with 7 h prior, generated the best prediction of RMSE = 9.113215. Predicting velocity values is at the highest accuracy of RMSE = 5.365601 with (p, d, q) of $(4, 1, 3)$ using 15 days of dataset with preceding 7 h., whereas predicting direction is at the highest accuracy of RMSE = 9.07044, with (p, d, q) of $(4, 0, 1)$ using 7 days of dataset with preceding 7 h. Nonetheless, we also found that, in some cases, when choosing the right configuration of (p, d, q) and historical range of datasets in practice, the validity of the estimated values can sometimes be less important than the accuracy aspect. For example, in the case of V comp prediction, although a combination of $(2, 0, 1)$ as (p, d, q), using 7 days of dataset with 7 h prior yielded the best prediction in terms of both RMSE (9.113215) and validity (all four estimates), the second best of RMSE (2.322) and validity (but only three out of four can be calculated) actually gave a better prediction when we added more sites (343 sites). There is a trade-off between RMSE and validity, and we recommend prioritizing obtaining a lower RMSE for better overall predictions. We also tested the model with higher degrees of p and q to see if they were affected in terms of the accuracy level and computational time. The result showed that even though the degrees of p and q were increased, the accuracy levels did not improve significantly. Figure 6 shows a sample result of increasing p (up to 20), which unfortunately did not improve accuracy performance. However, the increase also took a substantial amount of computational time to process.

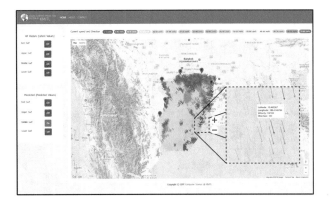

Fig. 5. The screenshot of our ocean surface current predictive system visualized on the map.

Table 9. An averaged RMSE of each method

Method	RMSE (cm/s)	
	U Comp	V Comp
Baseline [7]	7.02	20.43
LSTM [7]	5.85	21.2
ARIMA [7]	4.33	11.81
kNN [7]	4.41	7.58
Our method	10.79	6.23

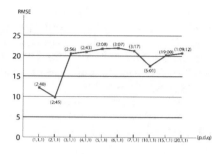

Fig. 6. Predicted performance with diff. p.

5 Conclusion and Future Work

In this research, the grid-based spatial ARIMA model is proposed and tested based on traditional ARIMA and dataset range selection criteria to analyze and predict direction and velocity values of the ocean surface currents in the Gulf of Thailand. To evaluate the performance, the prediction results from our method were compared with other methods, including baseline, kNN, ARIMA, LSTM. The results showed that our spatio-temporal data mining approach generated quite a high level of accuracy for V comp prediction. The method and techniques proposed in our approach can further be adopted and used for other GIS applications in which spatio-temporal forecast values are involved in analyzing ocean current patterns. Further adoption can maximize the positive impacts of our proposed prediction method. Such adoption could be facilitated by the diffusion of innovations theory, a communication framework for innovation and technology adoption [9]. The experimental results from this research showed the importance of selecting predictive data and the use of various parameters.

In future development, more potential/relevant parameters should be considered, such as increasing the number of surrounding points in grid for modeling. More related data sets can also be used, such as in the same period of the previous year (year-over-year) or the same season. Finally, we believe that the proposed model can generate better predictions to further guide crisis communication in real time via mass communication and social media for evacuation efforts to save lives in the event of coastal natural disasters.

References

1. Barrick, D.E., Evan, M.W., Weber, B.L., et al.: Ocean surface currents mapped by radar. Science **198**, 138–144 (1977). https://doi.org/10.1126/science.198.4313.138
2. Marta-Almeida, M., et al.: Efficient tools for marine operational forecast and oil spill tracking. Mar. Pollut. Bull. **71**, 139–151 (2013)
3. Cheng, T., Wang, J.: Applications of spatio-temporal data mining and knowledge discovery (STDMKD) for forest fire prevention. School of Geography and Urban Planning Sun Yat-sun University, Guangzhou, P R China (2008)

4. Shekhar, S., et al.: Spatiotemporal data mining: a computational perspective. ISPRS Int. J. Geo-Inf. (2015). https://doi.org/10.3390/ijgi4042306
5. Mamoulis, N., et al.: Mining, indexing, and querying historical spatiotemporal data. In: Proceedings of the ACM SIGKDD KDD 2004, pp. 236–245. ACM, New York (2004)
6. Aunthong, A.: Autoregressive integrated moving average model, Social Research Institute, Chiang Mai University (2007)
7. Jirakittayakorn, A., et al.: Temporal kNN for short-term ocean current prediction based on HF radar observations. In: Proceedings of the 2017 14th International Joint Conference on Computer Science and Software Engineering (JCSSE), pp. 1–6. IEEE Press, New York (2017)
8. Coastal Observing Research and Development Center, HFRNet National Network: WERA Radial LonLatUV (LLUV) file format. https://cordc.ucsd.edu/projects/mapping/documents/HFRNet_WERA_LonLatUV_RDL.pdf
9. Kee, K.F.: Adoption and diffusion. In: Scott, C., Lewis, L. (eds.) International Encyclopedia of Organizational Communication, pp. 41–54. Wiley-Blackwell, Hoboken (2017)

Parameter-Free Outlier Scoring Algorithm Using the Acute Angle Order Difference Distance

Pollaton Pumruckthum[(✉)], Somjai Boonsiri,
and Krung Sinapiromsaran

Department of Mathematics and Computer Science, Faculty of Science,
Chulalongkorn University, Bangkok 10330, Thailand
6071968923@student.chula.ac.th,
{Somjai.B,Krung.S}@chula.ac.th

Abstract. An anomaly scoring algorithm assigns the anomalous rating to an instance in a dataset which provides a large value for an outlier. In 2013, one of the parameter-free techniques called the Ordered Difference Distance Outlier Factor algorithm is proposed. It calculates a score using an ordered difference distance among all instances which derives from the distance matrix sorted in each row before computing the difference. The score contribution from other instances must be compared with the global minimum distance to avoid mis-detecting boundaries. However, this degrades the performance of the detection rate. To avoid the use of the global minimum distance term, the new technique is proposed using the ordered difference distance along the appropriate direction based on the acute angle. This technique is called the acute angle ordered difference distance outlier factor (AOF) algorithm. Three collections of ten synthesized datasets are designed to show the performance of AOF. The AOF algorithm reports very high scores for anomalies in synthetic datasets and has better performance than OOF when the anomalies are close together.

Keywords: Anomaly detection · Outlier factor · Ordered difference distance · Acute angle ordered difference distance

1 Introduction

Hawkins's definition for anomaly states that: "An outlier is an observation that deviates so much from other observations as to arouse suspicion that it was generated by a different mechanism" [1]. This is one of the cited definition of anomaly detection. An anomaly detection (also known as an outlier detection) refers to an identification algorithm to events or observations that do not conform to expected event or pattern [2]. It is one of the topics which has been studied in various fields. For example, in the stock market [3], an anomaly may be caused by the making of market manipulation so a stock player can avoid damage from it. In credit card transactions, anomalies can be caused by credit card fraud [4]. In time series, these anomalies in the data have a

© Springer Nature Switzerland AG 2020
P. Boonyopakorn et al. (Eds.): IC2IT 2019, AISC 936, pp. 37–45, 2020.
https://doi.org/10.1007/978-3-030-19861-9_4

negative impact on understanding the properties of the time series [5]. Moreover, building the forecasting model with anomalies will influence the model performance [6].

The anomaly scoring algorithm refers to the method of assigning scores to all instances in a dataset depending on the level of abnormality of each instance. It does not label any instance as an anomaly or a normal instance, as a matter of fact, it returns the high score of an instance to indicate that the instance has a high chance to be an anomaly. Many real-world problems normally apply the anomaly scoring algorithm than the anomaly detection due to the flexibility of setting up the threshold for detection anomalies.

One of the popular density-based scheme anomaly scoring technique is Local Outlier Factor (LOF) [7] that uses the k-nearest neighbors to compute the k-distance neighborhood for a multidimensional dataset. LOF helps to detect anomalies in a large dataset without assuming knowledge about the distribution of the dataset that makes it one of the popular anomaly scoring techniques.

Other density-based scheme is Connectivity-based Outlier Factor (COF) [8] algorithm. COF aims to improve the detection performance when a dataset has low density. This method computes score by the ratio of an average chaining distance of each instance compare with an average chaining distance of its neighbor.

Nonetheless, LOF and COF require the appropriate parameter setting. To avoid searching for optimal parameters, Order Distance Difference Outlier factor (OOF) [9] is proposed by Buthong et al. It uses the different distance between two instances to assign the score to each instance in the dataset. However, this technique has a flaw that it may use inappropriate different distance. This work, hence, uses an acute angle to select the appropriate instance in computing the different distance which helps generate a low score for an instance surrounding by other instances and a medium score for a boundary instance, and a high score for an anomaly.

2   Related Works

There are two popular approaches for anomaly scoring algorithms which are density-based and distance-based approaches.

2.1   Local Outlier Factor

In 2000, Breuning et al. proposed, Local Outlier Factor (LOF) [7], an anomaly scoring algorithm. LOF uses the k-nearest neighbors to find the neighborhood of each instance and then calculates the local reachability density of that instance and its neighborhood. Next, the LOF value calculated as the average ratio among all "local reachability densities" of neighborhood divided by the local reachability density of that instance.

2.2   Ordered Difference Distance Outlier Factor

In 2012, Bunthong et al. suggested the ordered distance difference Outlier Factor (OOF) [9], which uses the ordered difference distance from the computed instance which is the intended instance to assign the score. It is a parameter-free algorithm which initially generates the distance matrix of the dataset and arranges each row by its value to generate the ordered distance matrix. Next, OOF uses the ordered distance matrix to generate the difference distance matrix and calculates the score compared with the minimum distance from the whole dataset.

3   Main Work

3.1   Motivation

AOF aims to assign the score to an instance using the contribution from all other instances based on the distance. In this paper, an instance to be assigned the score is referred as *the computing instance*. A particular contribution from another instance that is assigned to the computing instance will be constructed. That instance is referred as *the reference instance* which may be blocked by the closest instance to the computing instance which is referred as *the covered instance*, see the middle figure of Fig. 1.

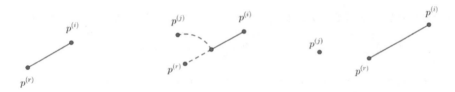

Fig. 1.   The AOF concept to assign a score to the computing instance from the reference instance that may be blocked by the covered instance

From Fig. 1, $p^{(i)}$ is the computing instance to be assigned a score to and $p^{(r)}$ is the reference instance to contribute a value to the score. On the left figure of Fig. 1 shows the direct contribution from $p^{(r)}$ to $p^{(i)}$ without any covered points. This happens when $p^{(r)}$ is the nearest neighbor of $p^{(i)}$ while the middle figure of Fig. 1 shows the contribution from $p^{(r)}$ to $p^{(i)}$ which is blocked by the covered instance $p^{(j)}$ using the difference distance from $p^{(i)}$ to $p^{(j)}$ with respect to $p^{(r)}$. Note that the angle from $p^{(r)}$ to $p^{(i)}$ and from $p^{(r)}$ to $p^{(j)}$ is acute. On the right figure of Fig. 1 $p^{(i)}$ and $p^{(r)}$ have no instance between them but other instances are closed to $p^{(r)}$ than $p^{(i)}$ in the opposite direction so the contribution will be the direct distance similar to the left figure of Fig. 1. Note that the angle from $p^{(r)}$ to $p^{(i)}$ and from $p^{(r)}$ to $p^{(j)}$ is obtuse. So all covered instances are ignored. After the contribution is accumulated from all the reference instances to the computing instance, then the AOF score is assigned as the average of all contributions.

3.2   Definition and Theorem

In this section, the background knowledge and definitions are introduced. Define D as a continuous dataset having n instances and m attributes. Define $d(p, q)$ as the Euclidean distance between p instance and q instance.

Definition 1: The distance matrix from D having n instances is the matrix by $\textbf{DistMtx} = (d_{a,b})_{n \times n}$ where $(d_{a,b})_{n \times n} = d(p^{(a)}, p^{(b)})$ for $p^{(a)}, p^{(b)} \in D$ and $a, b \in \{1, 2, \ldots, n\}$.

Definition 2: The ordered distance matrix of D is the matrix $\mathbf{O} = (d_{r,i_k}(r))_{n \times n}$ for $r \in \{1, 2, \ldots, n\}$ and $i_k^{(r)}$ is the index of the kth-ordered element in row r where $d_{r,i_1^{(r)}} \leq d_{r,i_2^{(r)}} \leq \cdots \leq d_{r,i_n^{(r)}}$.

This matrix O can be written as

$$O = \begin{bmatrix} 0 & d_{1,i_2^{(1)}} & d_{1,i_3^{(1)}} & \cdots & d_{1,i_n^{(1)}} \\ 0 & d_{2,i_2^{(2)}} & d_{2,i_3^{(2)}} & \cdots & d_{2,i_n^{(2)}} \\ \vdots & \vdots & \vdots & \ddots & \vdots \\ 0 & d_{n,i_2^{(n)}} & d_{n,i_3^{(n)}} & \cdots & d_{n,i_n^{(n)}} \end{bmatrix}$$

Note the first column of O contains only zero.

Definition 3: Given the computing instance $p^{(i)}$ in D and the reference instance $p^{(r)}$, $p^{(x)}$ is called the covered instance if the angle from $p^{(r)}$ to $p^{(i)}$ and from $p^{(r)}$ to $p^{(x)}$ is an acute angle while the distance from $p^{(r)}$ to $p^{(i)}$ is larger than the distance from $p^{(r)}$ to $p^{(x)}$.

The covered instance is the instance that will block the contribution from the reference instance to the computing instance. This blocks can be determined by the distance among these three instances as in Theorem 1.

Theorem 1: Given the computing instance $p^{(i)} \in D$ for $i \in \{1, 2, \ldots, n\}$ and the reference instance $p^{(r)}$ for $r \in \{1, 2, \ldots, n\} - \{i\}$. For $p^{(x)}, x \in \{1, 2, \ldots, i - 1\} - \{r\}$, if $(d_{i,x})^2 < (d_{r,x})^2 + (d_{r,i})^2$ then x is the covered instance.

This theorem helps finding all instances that are on the same side with the computing instance from reference instance. The AOF algorithm verifies the acute angle using the Pythagorean inequality theorem "a triangle is said to be an acute triangle if the square of the longest side is less than the sum of the squares of two smaller sides. In Fig. 2a, b, and c are the measures of each side of the triangle. Suppose c is the longest side, then $c^2 < a^2 + b^2$".

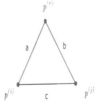

Fig. 2. The acute angle triangle.

Definition 4: Given $p^{(i)}$, $p^{(r)} \in D$ and $p^{(j)}$ is the covered instance of $p^{(i)}$ to $p^{(r)}$. The covered matrix of the dataset D is defined by $C = \left(d_{r,j_k^{(r)}} \right)_{n \times n}$ for $r \in \{1, 2, ..., n\}$ and $j_k^{(r)}$ is the covered instance for computing instance $i_k^{(r)}$ from O and C can be written as

$$C = \begin{bmatrix} 0 & 0 & d_{1,j_3^{(1)}} & \cdots & d_{1,j_n^{(1)}} \\ 0 & 0 & d_{2,j_3^{(2)}} & \cdots & d_{2,j_n^{(2)}} \\ \vdots & \vdots & \vdots & \ddots & \vdots \\ 0 & 0 & d_{n,j_3^{(n)}} & \cdots & d_{n,j_n^{(n)}} \end{bmatrix}$$

In other word, the covered matrix shows the distance between the reference instance $p^{(r)}$ and the covered instance $p^{(j)}$.

Note that the value in the first column is 0 since $p^{(r)}, p^{(i)}, p^{(j)}$ are the same instance and in the second column since $p^{(i)}$ is the closest instance to $p^{(r)}$ then there is no instance between those two instances. So, there is no covered instance for all $p^{(i)}$, therefore the distance value is 0.

Definition 5: Given $p^{(i)}, p^{(j)}, p^{(r)} \in D$ and $p^{(j)}$ is the covered instance of $p^{(i)}$ to $p^{(r)}$. The difference distance matrix by the acute angle of $p^{(i)}$ toward $p^{(r)}$ is defined by $\Delta AO = O - C$.

ΔAO is a distance between the computing instance and the reference instance if there is no covered instance otherwise it is a difference between the covered instance and the reference instance. For example, in the middle figure of Fig. 1. ΔAO is the difference between the distance between $p^{(r)}$ to $p^{(i)}$ and $p^{(r)}$ to $p^{(j)}$.

Definition 6: The Acute angle ordered difference distance Outlier Factor (AOF) is defined by $AOF(p) = \frac{\sum_{i=1}^{n} \Delta AO_{i,index(p)^i}}{n-1}$.

The score of the instance p is the average value of ΔAO which is the different distance to which they contribute to p with respect to other reference instances. If p is an outlier, then this average will be high while instances in the cluster which are surrounded by another instance will have a low value.

The following step is the AOF algorithm pseudo code.

```
Input    : The numerical dataset D with n instances
Step 1   : Compute the distance between all instances to
           generate the distance matrix
Step 2   : Order the value in every row in distance matrix
           to build ordered distance matrix O
Step 3   : Find the covered instance corresponding with
           acute angle of each instance to use in
           generating the covered matrix C
Step 4   : Generate the difference distance matrix ΔAO
Step 5   : Compute the AOF score for every instances
Step 6   : Order the instances according to their AOF value
Output   : The first n instances with ordered AOF
```

4   Experiment and Discussion

Figure 3 illustrates two examples of the AOF scores in one-cluster and two-cluster experiments. The left figure of Fig. 3 shows the dataset A with seven instances having six instances forming a cluster and one instance isolated from others. The right figure of Fig. 3 shows the dataset B with thirteen instances having two clusters containing six instances in each cluster while one instance is placed between two clusters. Figure 4 shows an example where two anomalies, $p^{(1)}$, and $p^{(2)}$, are close to each other, but they lie far away from other points.

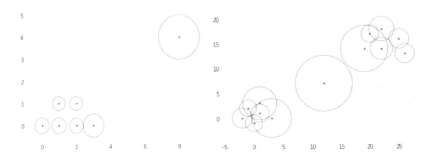

Fig. 3. AOF scores of instances with one-cluster dataset (Left) and with two-cluster dataset (Right)

Table 1. The OOF and AOF scores of the instance in dataset C

Instance	$p^{(1)}$	$p^{(2)}$	$p^{(3)}$	$p^{(4)}$	$p^{(5)}$	$p^{(6)}$	$p^{(7)}$	$p^{(8)}$	$p^{(9)}$	$p^{(10)}$
OOF score	1.01	0.90	0.57	0.61	0.44	0.53	0.32	0.28	0.24	0.31
AOF score	5.07	0.90	0.58	0.62	0.55	1.70	0.55	0.61	0.58	0.56

Table 1 shows the OOF scores comparing with AOF scores of all instances in the dataset C. It can be seen that the OOF score of $p^{(1)}$ and $p^{(2)}$ are very close to each other while the AOF score of $p^{(1)}$ is a lot larger than another points. The reason for instance $p^{(1)}$ having a quite different score between the OOF score and the AOF score is due to the minimum distance is used for both before assigning scores. On the other hand, the AOF is using the acute angle concept to compute the score instead of the global minimum. Therefore, the score for $p^{(1)}$ with other instances in the cluster as reference instance will be high, resulting in high score for instance $p^{(1)}$. When computing the AOF score for instance $p^{(6)}$ which is the border instance having no instance between $p^{(6)}$ and $p^{(1)}$ so the distance between these two instances is used to compute the score for $p^{(6)}$. This makes the score larger for the instance $p^{(6)}$ comparing with other instances in the cluster.

Three collections of ten synthetic datasets are used to measure the performance of the AOF algorithm. The ten datasets from the first collection contain 1010 instances where 1000 instances form a single cluster and 10 instances are anomalies. The ten datasets from the second collection contain 2020 instances where 1000 instances form

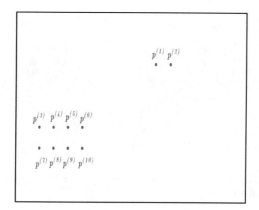

Fig. 4. Dataset C with two anomalies

the first cluster and other 1000 instances form the second cluster with 20 surrounding anomalies. The ten datasets from the third collection contain 1010 instances where 1000 instances form a single cluster and 10 instances form a pair of anomalies. Figure 5 shows the synthetic dataset from the collection 1 consisting of 10 synthetic datasets with the 1000 normal instances forming a single cluster and 10 anomaly instances. The radius surrounding each instance indicates the AOF score and the OOF score. So the instance with a large radius has a high probability to be an anomaly. Figure 6 shows the synthetic dataset from the collection 2 with 10 synthetic datasets with 2020 instances with two clusters. Each cluster has 1000 instances and 20 instances are anomalies. Figure 7 shows the result of collection 3 with 10 synthetic dataset with 1000 instances in the cluster and 10 anomalies which is close to each other in pair.

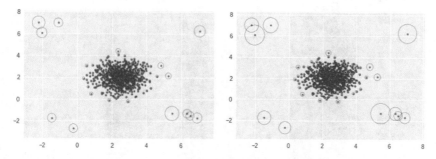

Fig. 5. The AOF scores (Left) and the OOF scores (Right) on the synthetic dataset with 1010 instances from the collection 1.

Figure 8 shows the average of the detection rate of the AOF and the OOF algorithms. When anomalies are far apart as in the first and the second collection, the detection rate of the AOF and the OOF algorithms are the same. However, in the third collection, when anomalies are close to one another, the detection rate of the AOF algorithms is much better than the OOF one due to the use of the global minimum

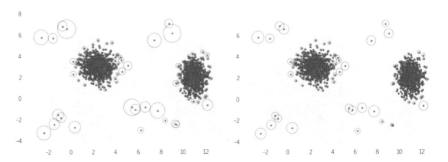

Fig. 6. The AOF scores (Left) and the OOF scores (Right) on the synthetic dataset with 2020 instances from the collection 2.

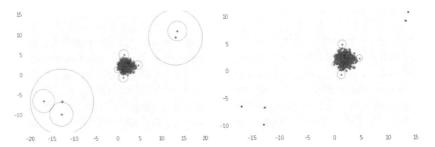

Fig. 7. The AOF scores (Left) and the OOF scores (Right) on the synthetic dataset with 1010 instances where 10 instances forming a pair of anomalies from the collection 3.

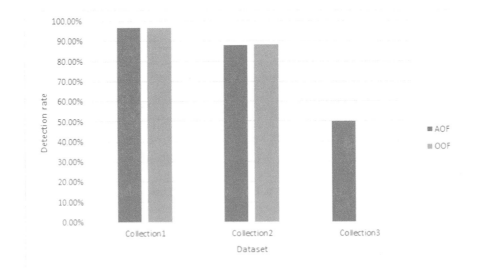

Fig. 8. The average of the detection rate of the AOF and the OOF algorithms for each collection

distance to assign a score while the AOF algorithm uses the angle concept for calculating the score so that the anomalies can be detected.

5 Conclusion

This paper proposes the AOF algorithm that calculates the score for each instance using the concept of acute angle to guarantee the appropriate contribution of other instances. The AOF score is calculated using the average of the distance contributions from all other instances in the dataset similar to the OOF score but it avoids to compare with the minimum distance as in the OOF algorithm, so the evaluation of the anomaly is different from the OOF score. A high AOF score shows that instance is far away from other instances, so this instance has a high probability of being an anomaly. Both the AOF and the OOF algorithms have a similar detection rate when anomalies are far from each other. However, the AOF algorithm gives a higher detection rate than the OOF algorithm when the anomalies are close to each other.

The anomaly score to detect datasets in the collection 3 can be improved further. The concept is to perform this anomaly score in separate round. Then extracting only the high AOF score to assign to the computing instance. Repeat this process without the assigned AOF score until the score is too low to be assigned anomaly.

References

1. Hawkins, D.M.: Identification of Outliers. Chapman and Hall, London (1980)
2. Chandola, V., Banerjee, A., Kumar, V.: Anomaly detection: a survey. ACM Comput. Surv. **41**(3), 15:1–15:58 (2009)
3. Golmohammadi, K., Zaiane, O.R.: Time series contextual anomaly detection for detecting market manipulation in stock market. In: IEEE International Conference on Data Science and Advanced Analytics (DSAA), Paris, pp. 1–10 (2015)
4. Yu, W., Wang, N.: Research on credit card fraud detection model based on distance sum. In: 2009 International Joint Conference on Artificial Intelligence, Hainan Island, pp. 353–356. IEEE (2009)
5. Basu, S., Meckesheimer, M.: Automatic outlier detection for time series: an application to sensor data. Knowl. Inf. Syst. **11**(2), 137–154 (2007)
6. Sagoolmuang, A., Sinapiromsaran, K.: Median-difference window subseries score for contextual anomaly on time series. In: 8th International Conference of Information and Communication Technology for Embedded Systems (IC-ICTES), Chonburi, pp. 1–6 (2017)
7. Breunig, M.M., Kriegel, H.P., Ng, R.T., Sander, J.: LOF: identifying density-based local outliers. ACM SIGMOD Record **29**(2), 93–104 (2000)
8. Tang, J., Chen, Z., Fu, A.W.C., Cheung, D.W.: Enhancing effectiveness of outlier detections for low density patterns. In: Chen, M.S., Yu, P.S., Liu, B. (eds.) Advances in Knowledge Discovery and Data Mining. PAKDD. Lecture Notes in Computer Science, vol 2336. Springer, Heidelberg (2002)
9. Buthong, N., Luangsodsai, A., Sinapiromsaran, K.: Outlier detection score based on ordered distance difference. In: International Computer Science and Engineering Conference (ICSEC), Nakorn Pathom, pp. 157–162 (2013)

Improved Weighted Least Square Radiometric Calibration Based Noise and Outlier Rejection by Adjacent Comparagraph and Brightness Transfer Function

Chanchai Techawatcharapaikul[1], Pradit Mittrapiyanurak[2], and Werapon Chiracharit[1(✉)]

[1] Department of Electronic and Telecommunication Engineering,
Faculty of Engineering, King Mongkut's University of Technology Thonburi,
Bangkok, Thailand
chanchai.oltai@mail.kmutt.ac.th,
werapon.chi@kmutt.ac.th
[2] Panasonic Research and Development Center Singapore, Singapore, Singapore
pradit.mittrayanuruk@sg.panasonic.com

Abstract. In this paper, we propose an improved radiometric calibration algorithm by extending the previous radiometric calibration method that is based on the "fixed exposure ratio" Brightness Transfer Function (BTF) and Weight Least Square. Our proposed method can be applied to the radiometric calibration on the case that the images are captured with non-constant exposure ratios which makes the image acquisition process to be more flexible. The key idea of our new proposed method is to determine a set of BTFs between the set of consecutive image pairs and then to form the set of composited functions. From the composition, we then convert the BTFs of all image pairs into the virtually single BTF. The noise & outlier rejection based on this BTF is applied to retrieve the clean data. Finally, a reformulation of weighted least square minimization is proposed to estimate the camera response function (CRF) of a camera. From the performance of our proposed algorithm in comparison with the state-of-the-art least square method proposed by Mitsunaga and Nayar as the baseline, we found that our method outperforms the baseline algorithm on both the synthetic dataset and the dataset of real-world images.

Keywords: Radiometric calibration · Noise and outlier rejection · Comparagrap · Brightness Transfer Function

1 Introduction

In digital camera, the Camera Response Function (CRF) is the mapping from sensor irradiance to image pixel intensity. Many computer vision algorithms assume that CRF must be linear, e.g., shape-from-shading and photometric stereo. Typically, the CRF of a digital camera is non-linear and unknown. Radiometric calibration is the process to estimate CRF from image observation. Once we know the CRF, we can transform the image in such a way that the linearity assumption is valid.

© Springer Nature Switzerland AG 2020
P. Boonyopakorn et al. (Eds.): IC2IT 2019, AISC 936, pp. 46–55, 2020.
https://doi.org/10.1007/978-3-030-19861-9_5

The most popular approach for estimating CRF is to use multiple images captured from a static scene with exposure time. The main approach [1–6] is to apply least square minimization on the error function that is based on the differences between the irradiances of corresponding pixels in multiple images (observations) captured at different exposure times.

In [7, 8], the authors address the shortcoming of least square minimization in which its estimation performance is weak when (i) the distributions of observations are non-uniform, and (ii) there are noise and/or outlier in the observations. To solve this shortcoming, the authors propose to use a rank minimization approach in the estimation of CRF.

In our previous work presented in [6], we propose a new radiometric calibration algorithm based on least-square minimization that can work more accurately than the work in [7, 8]. The key idea of our previous work in [6] is to use a weighted least square minimization with Brightness Transfer Function (BTF) based noise and outlier removal. However, there is a major shortcoming in our previous work [6] in which it works only under the assumption that the exposure ratio among the images acquired for calibration must be constant. The motivation of this paper is to extend our previous work as in [6] in such a way that it can work with the images whose exposure ratio are not necessary to be constant.

The key contributions of the work presented in this paper can be explicitly stated as follows. We propose a new radiometric calibration algorithm that is extended from [6] so that it does not need the requirement of constant exposure ratio anymore. Meanwhile, the accuracy of our new algorithm is still at the same level.

The rests of the paper are organized as follows. The related work is presented in Sect. 2. The new proposed algorithm is explained in Sect. 3. Next, the experiments with synthetic dataset and real world dataset are reported in Sect. 4. Finally, the conclusion is drawn in Sect. 5.

2 Related Work

One of the approaches that is widely used for radiometric calibration is to use a set of images of static scene taken with a fixed camera at varying exposure times for the CRF estimation. Our work presented in this paper is in the same line of this approach. In [1], the authors proposed the method to estimate the CRF which represented by a gamma-correcting function with the known of exposure ratio. In [2], a non-parametric method with smoothness constraint was proposed to estimate the CRF. Also, the authors assume that the exposure ratio is known. In [3], a least square minimization based method is proposed to estimate the CRF which is represented by a polynomial function. Unlike [2], the exposure ratio can be unknown and is simultaneously estimated at the same time that the CRF is estimated. Similarly, in [4], the authors proposed the method to simultaneously estimate the CRF and exposure ratio. However, unlike [3], the CRF is represented with a non-parametric representation. In [5], the authors proposed the method to estimate CRF from Brightness Transfer Function (BTF) which is computed from the comparagraph by dynamic programming technique. In [6], a weighted least square minimization in conjunction with BTF based noise and outlier rejection is proposed to estimate the CRF from the set of images of static scene taken with known exposure ratio. Unlike the above works that are mainly based on least square minimization, in [7, 8], the authors proposed to use a rank minimization method to estimate the CRF. In [9], the authors proposed the method to estimate the CRF based on BTF computed from the images. Later in [10], the authors proposed a low-parameter representation for CRF, referred to as EMoR (empirical model of response function), that was derived from a Database of real-world camera Response Functions (DoRF).

3   The Proposed Calibration Algorithm

In this algorithm, the pixel intensity observations from multiple images of static scene captured at various exposures are given as the input. If I denotes the irradiance and M denotes the pixel intensity, both terms are related by the CRF f by $M = f(I)$. The inverse CRF denotes $g() = f^{-1}()$ that means $I = g(M)$. We assume that the exposure ratios of the set of images pairs are known, however unlike our previous work [6], in this work we allow these exposure ratios to be different (not necessary to be equal). That is, if there are N images, $M_1, M_2, M_3, ..., M_N$, with the exposure times, $e_1, e_2, e_3, ..., e_N$, respectively. The exposure ratio of adjacent image pairs are defined as for example are $e_1/e_2 = \alpha_1, e_2/e_3 = \alpha_2, ..., e_i/e_{i+1} = \alpha_i, ..., e_{N-1}/e_N = \alpha_N$, respectively. The irradiance value at a pixel x of both adjacent images (M_i and M_{i+1}) are related by

$$g(M_i(x)) = \alpha_i g(M_{i+1}(x)) \tag{1}$$

where $M_i(x)$ is the intensity value of the image i at pixel x, $M_{i+1}(x)$ is the intensity value of the image $i + 1$ at the same pixel x. By considering all pixels $x \in X$ of all M images, the inverse CRF ($\hat{g}$) can be estimated by solving the weighted least square minimization expressed by the following equations.

$$\hat{g} = \arg \min_g \left(E_{WLS} + \lambda \sum_t H\left(-\frac{d(g(t))}{dt}\right) \right) \tag{2}$$

$$E_{WLS} = \sum_{i=1}^{n-1} \sum_{x \in X} (w_{i+1}(x)[g(M_i(x) - \alpha_i g(M_{i+1}(x))])^2 \tag{3}$$

$$w_{i+1}(x) = \frac{1}{1 + h(g(M_{i+1}(x)))} \tag{4}$$

where E_{WLS} is the formulation for the weighted least square error function. And $w_{i+1}(x)$ is the weight of pixel x which is computed according to the intensity of image $i + 1$ at pixel x. The $h(g(M_{i+1}(x)))$ is the number of pixel in the image $i + 1$ that their corresponding irradiance values are equal to $g(M_{i+1}(x))$. The monotonic increasing constraint is represented as $\frac{d(g(t))}{dt} > 0$ for all range of intensity values. And $H(\cdot)$ is the Heaviside step function in which $H(x) = 1$ for $x \geq 0$. Otherwise, $H(x) = 0$.

The derivatives $\frac{d(g(t))}{dt} > 0$ are computed by numerical differentiation at several equally sampled points. And λ is a large constant value to reflect the data that violate the monotonic constraint. To perform the minimization in (2), we represent the CRF with the polynomial representation in the same way as in [6] or [8]. We refer to [6] for more details.

However, the above minimization will not give good result when noise and outlier appear in images. Figure 1 shows an example of BTF on comparagraph with noisy & outlier. Note that when the exposure ratio of each image pair is constant.

In our previous work, we propose an algorithm to remove the noise and outlier based on the analysis of BTF. As mentioned previously, in this work, we allow the exposure ratio of each adjacent image pair not necessary to be equal. In case that the exposure ratio of each pair is equal with no noise and no outlier, the BTF can be shown in Fig. 1(b). That is we can fit this BTF with a parametric function. In the case of noise and outlier are present, we can apply RANSAC based Hermite Spline Fitting, as in [11], to remove the noise and outlier. However, if we allow the exposure ratio of each adjacent image pair to be different, the BTF plotted on the same axis can be shown as in Fig. 1(c). That is we cannot use one parametric function to represent the BTF.

To tackle the above limitation of non-constant exposure ratio, in this work we propose an extension of the brightness transfer function (BTF) based noise and outlier removal method presented in [6]. Specifically, this point is the contribution of our work presented in this paper. The detail of the extension can be explained as follows.

First, we build the set of BTFs from the consecutive image pairs. That is, for example if we have 5 images in the dataset, we create the BTF between the image 5 and the image 4, the BTF between the image 4 and the image 3, and the BTF between the image 3 and the image 2. In general, we define the function $h_{i,j}$ where, $i < j$, to represent the BTF between the ith image and the jth image. We assume that the exposure time of the ith image is less than the exposure time of the jth image. For example if we use 5 adjacent images, the BTFs are consist of 3 functions ($h_{4,5}$, $h_{3,4}$ and $h_{2,3}$).

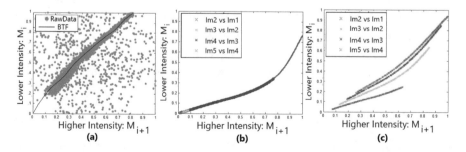

Fig. 1. Example of raw data as scatter plots in uniform distribution (a) noisy & outlier data with BTF, (b) clean 5 adjacent data with unique exposure ratio and (c) clean 5 adjacent data with various exposure ratio

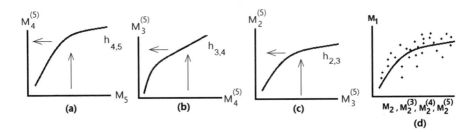

Fig. 2. Example of transformation from M_5 to M_2 and the last comparagraph

Then, we apply the pre-selection based on comparagram thresholding and the hermite spline fitting with RANSAC as presented in [6] to each pair of the data to estimate the BTFs (e.g., $h_{4,5}$, $h_{3,4}$ and $h_{2,3}$). As by product of the above procedure, we can partially remove the outlier and noise from each pair.

Second, from the BTFs we obtained in the previous step, we use these BTFs to transform the intensities of the images into so that we can plot the intensity pairs of all images in the axes of the BTF between the 1st image and the 2nd image.

The transformation procedure can be expressed with the following composite function:

$$M_k^{(L)} = h_{k,k+1}\left(h_{k+1,k+2}(\dots h_{L-2,L-1}(h_{L-1,L}(M_L)))\right) \tag{5}$$

where $M_k^{(L)}$ is intensity value that transformed the image L (M_L) to the intensity range of image k as if it is virtually acquired at the exposure time while we acquire the image k.

Intuitively, the meaning of (5) is that from the intensity acquired at the exposure time of the image L, we apply $h_{L-1,L}$ so that we will get a new image as if it is acquired at the exposure time of the image $L - 1$. We apply this series of transformations until we get the final image $M_k^{(L)}$ as if it is acquired at the exposure time of the image k.

Alternatively, we can explain the meaning of the composited transformation in (5) with an example of transformation from image 5 to image 2, i.e., $M_2^{(5)}$, with the help of Fig. 2 as follows. First, we transform the image 5 to image 4 by using the pre-calculate of BTF between image 4 and the image 5. This first step is representation by $M_4^{(5)} = h_{4,5}(M_5)$ as shown in Fig. 2(a). After that, from the transformed image $M_4^{(5)}$, we apply the same procedure with the function h_{34} as shown in Fig. 2(b), then we get the transformed image $M_3^{(5)} = h_{3,4}\left(M_4^{(5)}\right) = h_{3,4}(h_{4,5}(M_5))$.

We repeat the composited transformation so that we get $M_3^{(5)} = h_{3,4}\left(M_4^{(5)}\right) = h_{3,4}(h_{4,5}(M_5))$ as shown in Fig. 2(c) and finally $M_2^{(5)} = h_{2,3}\left(h_{3,4}(h_{4,5}(M_5))\right)$.

Finally, the comparagraph between image 1 and all transformed images $M_2, M_2^{(3)}$, $M_2^{(4)}$ and $M_2^{(5)}$ are created as in Fig. 2(d). For the purpose of illustration, we show these data as points in the figure. In the last step, we apply the BTF based noise and outlier removal as in our previous work.

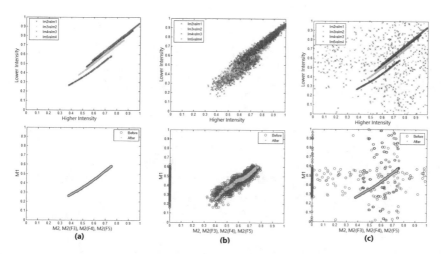

Fig. 3. Examples of comparative scatter plot in three types of data (a) Clean Data, (b) Noisy Data and (c) Outlier Data, Upper Row: Raw data without remove noise & outlier, Lower Row: Last comparagrap before (observation pixel: red pixel) and after (remained data: green pixel) remove noise & outlier data

The by-product of this step, we can select only the clean data to be used in the CRF estimation with the weighted least square minimization explained in Eqs. (2), (3), and (4). The more details of solving the minimization can be seen in our previous work [6].

It is important to emphasize that we trace back to the original data, e.g., M_5, M_4 and M_3, that are corresponding to the clean data in the transformed values, i.e. $M_2^{(5)}, M_2^{(4)}$ and $M_2^{(3)}$, respectively. These selected original data will be used in the minimization.

Note that, the reason that we use the original data in the minimization is to make the intensity range involved in the minimization to cover the whole range as much as possible.

In Fig. 3, we show the examples of our algorithm on three different cases of synthetic datasets: (a) clean data case, (b) noisy data case, and (c) outlier data case.

The figures in the upper row of Fig. 3. are the figures of the scatter-plots before applying the composited transformation explained above. As the exposure ratios of these pairs are non-constant, we can see that the data are not on the same BTF. Meanwhile, after apply our proposed algorithm of the computation of the composited transformation, the data will lie on the same BTF as if they are virtually acquired with the same exposure ratio. Therefore, we can decide whether a data point is clean data (shown with green color in Fig. 3 bottom row) or noise/outlier data (shown with red color in Fig. 3 bottom row).

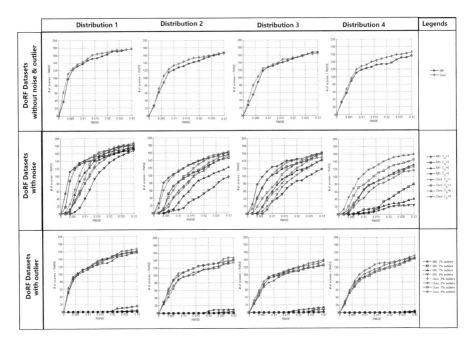

Fig. 4. Results on synthetic DoRF datasets.

Additionally, we can see the motivation of our proposed algorithm in the figures of the bottom row of Fig. 3. That is, after applying the composited transformations, it can bring all data that are acquired with non-fixed exposure ratios into the data that can represented with the single BTF. From this BTF, we can consider the data that are deviated from the BTF as noise and outlier. Therefore we can keep only the clean data to use in the CRF estimation.

4   Experiments

The proposed algorithm is evaluated on both synthetic dataset and real-world image dataset. The results of our CRF estimation are compared with the baseline method proposed by [3]. Similar to [6], we use the 6th degree polynomial function to represent the CRF and solving the minimization in (2) is initialized with the linear CRF.

For the experiment on synthetic data, we use the DoRF dataset [10] which consist of 201 CRFs. We generate the synthetic data in the same way as in [8] which can be explained as follows. First, we generate the scene radiances in the range of [0, 1] with four different distributions referred to as D1 (Uniform), D2 (Multiple peaks), D3 (Multiple peaks with negative skew), and D4 (Multiple peaks with positive skew). Due to page limitation, we refer the reader to see the pictures of these distribution in [6].

To simulate the multi-exposure data of each test case, five irradiance observations of size 1,000 (1,000-pixel irradiance image) are generated from a scene radiance with five different exposure times which are 1/5, 2/5, 3/5, 4/5 and 1, respectively. Remark that, the exposure ratios between the consecutive image pairs are not constant.

We evaluate the effectiveness of our proposed algorithm against noise, we add poisson noises with four different camera gains (C_g = 1, 3, 6 and 9). Also, we evaluate the robustness to outlier by randomly changing selected pixel values with 4 different levels (3%, 5%, 7%, and 9%) to the irradiance observations.

The performance of radiometric calibration algorithms are calculated by the Root Mean Square Errors (RMSEs) that is computed from the error between the CRF obtained from an algorithm and the groundtruth CRF. We consider that a calibration (CRF estimation) is successful if the corresponding RMSE being less than the threshold (in this paper use 0.03). Then we compute the cumulative histograms of number of successful results.

Based on the evaluation procedure explained previously, the results on DORF dataset can be seen in Fig. 4. From the results, we can see that our algorithm outperforms the baseline method [3] both in the case of data with noise and outlier. That is, the number of successful calibrations from our method is greater than the one of baseline method. Additionally, we can observe from the result that in case of data without noise & outlier (see Fig. 4 top row), our algorithm gives the successful rate almost equal to the baseline method. However, in case of data with noise (see Fig. 4 middle row), the bias in observation data (the distributions 2, 3, and 4) have an impact on the results. That is, it significantly degrades the successful rate of the baseline method due to the over fitting problem. Meanwhile, our algorithm (by using weighted least square minimization) can tolerate the bias effect as we can see that the number of successful results are greater. For the case of data with outlier (see Fig. 4 bottom row), we can see that the number of successful results obtained from our algorithm is very close to the results on the clean data. This result indicates that our noise & outlier rejection module can effectively handle the outlier in data. Also, we can see that our algorithm are robust to various degrees of outliers, i.e., the number of successful results on different levels of outliers are almost the same.

For the experiment on real-world images, we perform the experiment on the images from three different datasets: (i) outdoor scene dataset used in [7], (ii) indoor scene dataset used in [7], and (iii) color chart dataset in [12]. The examples of images in these datasets can be seen in Fig. 5, leftmost column. The images in each dataset consists of 5 images of static scene acquired at different known exposure times. Note that the exposure ratios of consecutive image pairs are not constant. These 5 images will be used for CRF estimation. Additionally, in each dataset, there are one RGB test image accompanied with the groundtruth RAW retrieved from camera.

Once we estimate the CRF from 5 images for each dataset, we apply the inverse CRF to the test RGB image to obtain the calibrated RGB image. Similar to [7], we consider that the white-balance and color space transformations are included in the pipeline of converting RAW to RGB image. We consider that these transformations are represented with a single 3×3 matrix. This matrix is estimated by applying a least square method to the corresponding pixel intensities between the groundtruth RAW and the calibrated RGB image. Once we get this 3×3 matrix, we use the transformation matrix to convert the calibrated RGB image into the estimated RAW image. Then, we calculate the error map (or difference image) between the groundtruth RAW image and the estimated RAW image.

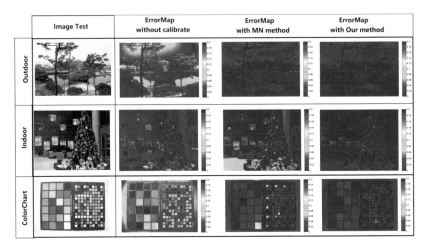

Fig. 5. Results on real world test. The images show pseudo-color error map for the green channel: (i) without using calibration, (ii) with the MN calibration method, and (iii) with our calibration method.

The error maps on the test images of 3 datasets are shown in Fig. 5. The error maps on the 3rd and 4th columns are obtained from the baseline radiometric calibration method and our proposed method, respectively. Meanwhile, the error maps on the second column are obtained when we apply the 3×3 matrix without radiometric calibration. Note that, due to page limitation, the above error maps are computed and shown from the green images (G channel) only. In Table 1, we show the quantitative results of error map in terms of RMSEs on each R, G, B channels.

From the results in Fig. 5 and Table 1, our proposed algorithm outperforms the baseline algorithm. We can observe qualitatively in the error maps shown in Fig. 5 that our method (4th column) is better than the baseline method (3rd column), especially on the indoor scene dataset and the color chart dataset.

Table 1. RMSE in each page of Real world Test (The minimum value of each page is bold)

Image test	RMSE in each page of Realworld Test								
	Outdoor			Indoor			Color Chart		
	R	G	B	R	G	B	R	G	B
W/O	0.0250	0.0540	0.0394	0.0545	0.0541	0.0281	0.0783	0.0239	0.0204
MN	0.0099	0.0201	0.0139	0.0487	0.0705	0.0342	0.1310	0.0728	0.0189
OURs	**0.0075**	**0.0142**	**0.0101**	**0.0294**	**0.0229**	**0.0128**	**0.0531**	**0.0229**	**0.0113**

5   Conclusion

In this paper, we propose an improved radiometric calibration algorithm by extending the previous work presented in [6]. This algorithm can work with the images whose exposure ratio is not necessary to be constant. The main idea of the proposed method is to adapt the adjacent comparagraph and BTF for noise and outlier rejection.

Also, the weighted least square minimization is reformulated for supporting non-constant exposure ratios. The evaluation on both synthetic data and the real-world images has shown the promising results of our proposed algorithm. Some of possible future works can be listed as follows. First, we can use a better optimization method to avoid the local minimum problem (not converge to correct solution). Second, we can search for a more rigorous approach of weight definition in the estimation.

Acknowledgment. This work is supported by "Thailand Graduate Institute of Science and Technology (TGIST)" funded under "National Science and Technology Development Agency (NSTDA)", Thailand. The first author is grateful to Dr. Pakorn Kaewtrakulpong who is a former Associate Professor at KMUTT, Thailand and Dr. Supakorn Siddhichai who is a former research scientist at NECTEC, Thailand.

References

1. Mana, S., Picard, R.W.: On being 'undigital' with digital cameras: extending dynamic range by combining differently exposed pictures. In: Proceedings of IS&T's 48th Annual Conference, pp. 422–428 (1995)
2. Debevec, P.E., Malik, J.: Recovering high dynamic range radiance maps from photographs. In: Proceedings of the 24th Annual Conference on Computer Graphics and Interactive Techniques, SIGGRAPH 1997, New York, NY, USA, pp. 369–378 (1997)
3. Mitsunaga, T., Nayar, S.: Radiometric self calibration. In: Proceedings of IEEE Conference on CVPR, vol. 1 (1999)
4. Mann, S., Mann, R.: Quantigraphic imaging: estimating the camera response and exposures from differently exposed images. In: Proceedings of IEEE Conference on CVPR, pp. 842–849 (2001)
5. Kim, S.J., Pollefeys, M.: Robust radiometric calibration and vignetting correction. IEEE Trans. PAMI 30(4), 562–576 (2008)
6. Techawatcharapaikul, C., et al.: Improved radiometric calibration by brightness transfer function based noise & outlier removal and weighted least square minimization. IEICE Trans. Inf. Syst. **E101-D**(8), 2101–2114 (2018)
7. Lee, J.Y., et al.: Radiometric calibration by transform invariant low-rank structure. In: Proceedings of IEEE Conference on CVPR, pp. 2337–2344 (2011)
8. Lee, J.Y., et al.: Radiometric calibration by rank minimization. IEEE Trans. PAMI 35(1), 144–156 (2013)
9. Grossberg, M., Nayar, S.K.: Determining the camera response from image: what is knowable? IEEE Trans. PAMI 25(11), 1455–1467 (2003)
10. Grossberg, M., Nayar, S.K.: Modeling the space of camera response functions. IEEE Trans. PAMI 26(10), 1272–1282 (2004)
11. Fischler, M.A., Bolles, R.C.: Random sample consensus: a paradigm for model fitting with applications to image analysis and automated cartography. Commun. ACM 24(6), 381–395 (1981)
12. Kim, S.J., et al.: A new in-camera imaging model for color computer vision and its application. IEEE Trans. PAMI 34(12), 2289–2302 (2012)

Application of Data Mining, Language, and Text Processing

Fuzzy TF-IDF Weighting in Synonym for Diabetes Question and Answers

Ketsara Phetkrachang and Nichnan Kittiphattanabawon[✉]

School of Informatics, Walailak University, Nakhon Si Thammarat, Thailand
ketsara.p@hotmail.com, knichcha@mail.wu.ac.th

Abstract. Currently, the synonyms are a problem to retrieved answer from question answering systems. A fuzzy based similarity is one method that many researchers used to solve this problem. This paper applied the fuzzy method with TF-IDF weighting in considering the alphabet of words in order to analysis a similarity between words. Our corpus consists of five hundred answers collected from reliable medical resources. Several fuzzy conditions were investigated to find out the best condition for answering the question. To evaluate our proposed method, thirty frequently asked questions are tested and compared to experts answers. The results showed that the acceptable answers were discovered on 80% words similarity above (fuzzy degree is greater than 0.8) with 80.09% or more of precision.

Keywords: Question and answer · Diabetes · Fuzzy

1 Introduction

One of the challenges of the questions answering system is to provide the correct answer for the user's question. The question answering systems have special significance and advantages over search engines and are considered to be the ultimate goal for user's information needs [1]. A question answering systems still have some problems in terms of finding answers because some words have similar meanings. The system will try to understanding and analyze a part of message that could be the answer. Over the years, many researchers have studied the question answering system. Most answers are based on approach a semantic analysis of the question [2, 3]. A fuzzy based similarity is done by means of a fuzzy matching keyword between keywords in questioning and keywords in the answers, which returns a list of results based on likely relevance even though search words and spellings may not exactly match, for example: Irfan et al. [4] to implementation of Fuzzy C-Means algorithm and TF-IDF on English journal summary. They use TF-IDF on a set of documents. Then a cosine distance matrix is constructed and hierarchical agglomerative clustering and fuzzy K-means algorithm is applied to get the clusters. Romero et al. [5] proposed the development of a Clinical Decision Support System framework to diagnose a set of fuzzy diseases, concretely applied to Fibromyalgia and associated syndromes. For this purpose, in this paper a reasoning method that uses theories about conceptual categorization from the psychology, pattern recognition, and Zadeh's prototypes has been designed. Through the use of this model, satisfactory results in the evaluation of patients were obtained.

© Springer Nature Switzerland AG 2020
P. Boonyopakorn et al. (Eds.): IC2IT 2019, AISC 936, pp. 59–68, 2020.
https://doi.org/10.1007/978-3-030-19861-9_6

Falomir et al. [6] proposed an approach for scene understanding based on qualitative descriptors, domain knowledge and logics. In [7] Sappagh et al. presented a case-based reasoning framework based on fuzzy ontology for diabetes diagnosis. A fuzzy semantic retrieval algorithm was built for case retrieval. They also implemented a decision support system to support the decision maker in the diagnosing process. Research on question answering systems has been continuously developed. To solve problems and optimize the choice of answers and apply them to the appropriate theory. Abacha et al. [8] proposed types of questions for medical QA system focus two main types: (1) factual questions expressed by WH pronouns and (2) boolean questions expecting a yes/no answer. An answer can be (1) a medical entity for factual questions or (2) Yes or No for boolean questions. Terol et al. [9] presented a logic based medical question system that is intended to handle the 10 common question types applied to medical disciplines in modern QA over restricted domains. Wu et al. [10] presented mining query subtopics from questions in the community question answering. The subtopics are represented as a number of clusters of questions with keywords summarizing the clusters. They proposed method significantly outperforms the existing methods in terms of keyword extraction while achieving a comparable performance to the state-of-the-art methods for question clustering. Weis et al. [11] reported to extend the existing LogAnswer solution by a casebased reasoning (CBR) approach, where annotated answer candidates for known questions provide evidence for validating answer candidates for new questions.

To represent term importance in documents for text processing, it is necessary to weight the terms in the documents. There are several ways in weighting terms. TF (term frequency), TF-IDF (term frequency-inverse document frequency) [12]. TF-IDF weighting is applied to assess how important a word is with respect to a document in a corpus. The importance of word grows proportionally to the number of times a word appears in the document, but offsetted by the frequency of the word in the collection. Variations of the TF-IDF weighting scheme are often used in considering a document's relevance given by a user query. TF and IDF are the two criteria to measure the weight of a word within text, The TF-IDF weighting schema assigns to term a weight in the document and the term weighting schemes.

TFIDF is evolved from IDF which is proposed by Salton with the heuristic intuition that a query term which occurs in many documents is not a good discriminator, and should be given less weight than one which occurs in few documents. Equation (1) is the classical formula of TFIDF used for term weighting.

$$W_{jk} \; = \; tf_{ik} \; \times \; idf_j \tag{1}$$

where w_{jk} is the weight of term j in document k, tf_{jk} is the number of term j that appears in document k, and idf_j is the inverse document frequency of term j as derived in the equation

$$idf_i = log\frac{N}{df_j} \tag{2}$$

In this paper, we compare the effectiveness of answer using fuzzy method with TF-IDF weighting in considering the alphabet of words in order to analysis a similarity between words.

The rest of the paper is organized as follows. Section 2 proposed methods, including the design framework for question answering system in diabetes care. The next section expresses details of experimental results and discussions. Finally, the conclusion is made in the last section.

2 Proposed Method

This section presents a method to select corresponding answers using Fuzzy TF-IDF weighting in Synonym for Diabetes Question and Answers. The conceptual framework for the diabetes care system consists of three main steps, i.e., (1) question and answers pre-processing, (2) fuzzy answer scoring, and (3) question-answer matching, as shown in Fig. 1.

2.1 Question and Answers Pre-processing

We conducted the research on 30 questions and 500 suggestions. The questions and answers were collected from FAQ (frequency asked question) section of several Medical websites, e.g., Endocrine Society of Thailand (www.thaiendocrine.org), Diabetes Association of Thailand (www.diabassocthai.org), and Thai Health Promotion Foundation (www.thaihealth.or.th), and Moh-chao-Ban Magazine (www.doctor.or.th). Both 30 questions and 500 answers were prepared by pre-processing steps. Firstly, each of questions and each of answers were segmented in a keyword segmentation step corresponding terms in a predefined dictionary. There are 76,714 terms in the dictionary including general words and medical terms. Secondly, stop word elimination step was processed to the segmented words. The segmented words which matched to the stop words would be removed from the consideration. We have 1,879 stop words in a collection which contains the most common words in this area. Finally, the questions and answers were separated in to terms without stop words.

2.2 Fuzzy Answering Scoring

This step is to measure the similarities between question and answer by using Dice coefficient [13] of question (Q) and Answer (A). Assigned q and A is vector; q is question, A is document of answer, x is counting the characters shared by both (Q and A), y is counting the characters occurring in question (Q) only, z is counting the characters occurring in answer (A) only then the result will be calculated by the following equation.

$$\text{Score} = \frac{2x}{2x+y+z} \tag{3}$$

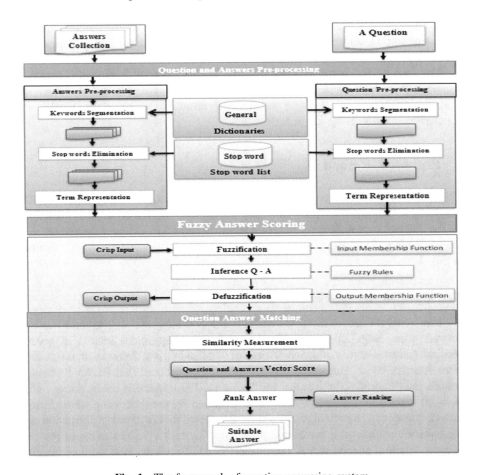

Fig. 1. The framework of question answering system

The process of fuzzy reasoning is incorporated into what is called a fuzzy inferencing question and answers. It is comprised of three steps that process the system inputs to the appropriate system outputs. These steps are (1) Fuzzification, (2) Inference Question-Answering and (3) Defuzzification. Degree of membership: degree to which a crisp value is compatible to a membership function, value from 0 to 1, also known as truth value or fuzzy input. membership function, MF: defines a fuzzy set by mapping crisp values from its domain to the sets associated degree of membership. crisp inputs: distinct or exact inputs to a certain system variable, usually similarities that can be measured such as Fuzzy degree ≥ 0.8 mean the finding of the familiar of word in query and answer by considering only some words that appearance in question and answer and will result more than 08.

2.3   Question Answering Matching

This procedure is the matching between question and answer.

Similarity Measurement
This step is to measure the similarities between question and answer by using cosine similarities [14], which define q and A are vector. Assigned q and A is vector; q is question, A is document of answer, i is sequence of answer then the result will be calculating by the following equation.

$$\text{Sim}(q, A_i) = \frac{q \times A}{[q] \times [A]} \tag{4}$$

Rank Answer
It is the sorting of the answer step, which is determined from the result sorting result in descending order.

3 Experiments

3.1 Experimental Setup

We conducted the experiments on 30 questions and 500 suggestions. The example of question in corpus such as "ถ้ามีอาการใจสั่น มือสั่น หิว ทำอย่างไร?" (How to do if found the symptom of hand shake, heart palpitate, hungry?). The experiments were divided into 4 comparisons, i.e., (1) an overall performance among fuzzy degrees, (2) precision and recall among similarity scores and fuzzy degrees, (3) the number of correct answers to the experts' answers, and (4) the performance of our method group by question lengths and question types. Fuzzy degree of number consists of 0.5 to 1.0 (fuzzy of 0.5 means 50% the same writing of a term of a question and that of an answer), (fuzzy of 0.6 means 60% the same writing of a term of a question and that of an answer), (fuzzy of 0.7 means 70% the same writing of a term of a question and that of an answer), (fuzzy of 0.8 means 80% the same writing of a term of a question and that of an answer), (fuzzy of 0.9 means 90% the same writing of a term of a question and that of an answer) and (fuzzy of 1.0 means 100% the same writing of a term of a question and that of an answer). To evaluate our proposed framework, we have three experts to check answers which are suggested from the system. If two of the experts agree on the result, they will be considered as a correct suggestion. If there is only one or no one coincides with the result, the suggested answer will be accounted as a wrong suggestion. The performance of the whole system will be assessed by precision, recall and F-measure.

3.2 Experimental Results and Discussions

Comparing Performance of Fuzzy Degrees
The first experiment showed a performance of fuzzy value in various fuzzy degrees. We chose specified fuzzy degrees from 0.5 to 1.0 since the degrees below 0.5 normally

did n't performed well [15, 16]. Precision, recall and F-measure were evaluated among top-N, which N = 1 to 5. Here, top-N means N first answers which have the highest similarity score. Table 1 presented the percentage in precision, recall, and F-measure on fuzzy degree of 0.5–1.0, varied by top-N. Table 2 presented the performance on type and size of question.

To compare the precision and recall among fuzzy degrees, we plotted them to a line graph as shown in Fig. 2. The graph showed the performance from fuzzy degree of 0.5 to 1.0. The horizontal axis on the graph represented the N value in top-N and the vertical axis represented the percentage of precision or recall. For all degrees, the graph presented the decline for the recall and the increase for the precision. From N is 1 to 2, there was a decline in precision value and an increase in recall value dramatically, for example, as shown in Table 1, 92.50% downto 48.70% of precision and 61.89% upto 73.11% of recall on 1.0 fuzzy degree, 86.39% downto 46.94% of precision and 60.78% upto 74.22% of recall on 0.9 fuzzy degree, and 80.09% downto 49.25% of precision and 54.94% upto 69.78% of recall on 0.8 fuzzy degree.

The results of the experiment found clear support for the best precision of top-1 since the precision at top-1 was the most outstanding for all fuzzy degrees. It could simply mean that the best answers are the answers which have the highest similarity score. Moreover, This is an important finding in the understanding of the performing of fuzzy degrees. Unlike logical co-operators method which must compare the exactly term spelling [17] a fuzzy method is not necessary to have the same term spelling. Fuzzy degree of 0.8 above (0.8–1.0) performed well, giving good results on top-1. This shows that the similarity of a term in questions and answers is not necessary be the be the same (fuzzy of 1.0 means exactly the same writing of a term of a question and that of an answer), (fuzzy of 0.9 means 90% the same writing of a term of a question and that of an answer), and etc. Another promising finding was that the recall is better when the N in top-N increased. This means that all correct answers will be covered when the similarity score decreased. The reason is that there are more than one correct answers for any questions. However, our method can retrieve some correct answers at the top-1 with the good precision. Note that when the recall is better, the precision will be worst.

Comparing Performance of Similarity Scores

To clearly confirm the first experiment, this experiment presented similarity scores for each fuzzy degree. The similarity scores were represented by top-N, where N is 1 to 5. The values of N were put on the bars. The less value of N, the more value of similarity score, e.g., top-1 means that the answers derived from our method had the highest score of similarity, on the other hand, top-5 identified that the answers got from our method had the lowest score of similarity. Figure 3 showed the bar chart of precision and recall varied by fuzzy degrees. For each degree, we compared the performance among top-N, which implicitly means a level of similarity score as described above. It can be seen that the precision of each fuzzy degree fell suddenly from top-1 to top-2, but fell gradually from top-2 to top-5. Conversely, the recall of each fuzzy degree grew a little bit steeply from top-1 to top-2, but went up steadily from top-2 to top-5. Moreover, both precision and recall are gently rose along the fuzzy degrees. The findings are directly in line with previous findings. The results explicitly demonstrated that our method yields increasingly good results on top-1 for all fuzzy degrees, especially 0.8, 0.9 and 1.0 of fuzzy degrees which had the precision that was greater than 80 percent. The results confirmed that this is a good choice for giving the suggestions for the specified question. Extensive results carried out showed that the recall was better when the value N in top-N increased. A similar conclusion was reached by the number of correct answers since there are more than one correct answer from the experts. As some answers were good while some of them might be not quite good, therefore, the bad answers could be in the top of lower N (top-2, top-3, top-4, and so on) since they had the less similarity score.

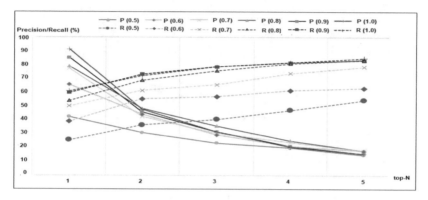

Fig. 2. The comparison between a precision and recall on fuzzy degree of 0.5–1.0 varied by top-N

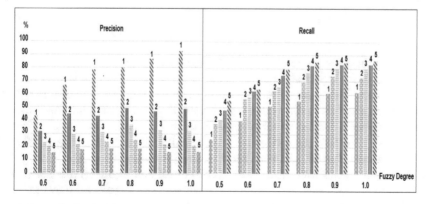

Fig. 3. The precision and recall on top-N (similarity score) varied by fuzzy degrees

Table 1. The performance in precision, recall, and F-measure on fuzzy degree of 0.5–1.0 varied by top-N

top-N	Precision	Recall	F-measure	top-N	Precision	Recall	F-measure
Fuzzy degree = 0.5				Fuzzy degree = 0.6			
top-1	43.33	26.22	30.44	top-1	66.67	30.89	45.94
top-2	31.67	37.31	31.35	top-2	45.11	45.25	43.94
top-3	24.04	41.31	27.02	top-3	30.05	57.92	34.11
top-4	20.53	47.97	25.46	top-4	21.64	62.36	27.09
top-5	15.46	55.31	21.31	top-5	18.28	64.03	24.42
Fuzzy degree = 0.7				Fuzzy degree ≥ 0.8			
top-1	78.33	51.08	56.63	top-1	80.09	54.94	59.70
top-2	43.46	62.42	42.60	top-2	49.25	69.78	48.49
top-3	31.28	66.53	36.05	top-3	36.59	76.67	42.22
top-4	24.50	74.72	31.52	top-4	25.66	81.94	33.85
top-5	18.44	79.44	25.88	top-5	18.35	84.72	26.22
Fuzzy degree = 0.9				Fuzzy degree = 1.0			
top-1	86.39	60.78	65.32	top-1	92.50	61.89	61.91
top-2	46.94	74.22	47.75	top-2	48.70	73.11	49.01
top-3	32.30	79.58	38.84	top-3	32.28	79.58	38.82
top-4	21.67	82.75	28.87	top-4	21.10	82.75	29.43
top-5	16.30	84.42	24.16	top-5	16.27	85.94	24.18

Table 2. The performance on types and sizes of questions

Type	Total	Descriptive			Exact				
Size	All	Why	How	Sub total	What	When	Where	Yes/no	Sub total
2–3 term	14(46.67%)	1	6	7(23.33%)	2	0	0	5	7(23.33)%
4–6 term	16(53.33%)	2	1	9(30.00%)	1	2	1	3	7(23.33)%
All	30(100.00%)	3	7	16(53.33%)	2	2	1	8	14(46.66%)

Comparing the Number of Correct Answers to the Experts

The two previous experiments mentioned that the recall improved when the value N in top-N increased. This experiment intended to show why the recall increased while the precision decreased as the fuzzy degrees went up. Figure 4 showed the comparison of number of answers derived from both experts and our method. There are three lines in the chart. All three lines represented the average number of answers on top-N, where N is 1 to 5. The grey line with diamond showed the correct answers by experts, the yellow line with square dealt with the answers (both correct and incorrect) that our method suggested to the questions. and the green line with triangle represented only the correct answers from our method. The graph reported that the average number of answers by experts was 2.67 while those retrieve by our method were 0.14 for the values of N were from 1 to 5, respectively. The result now provided evidence to the recall. Since the average correct numbers (by experts) was 2.67, our result in top-1 had the average number of 1.20.

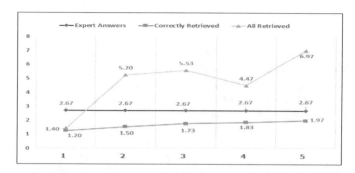

Fig. 4. The average number of answers comparing to experts

Comparing the Performance by Question Types and Question Sizes

The last experiment investigated the performance on types and sizes of questions. For types of question, we divided the questions to 2 categories, i.e., descriptive and exact.

The descriptive type is the questions in the group of why-question and how-question, which will have the long answers. What-question, when-question, where-question and yes/no-question were grouped to the exact type. As sizes of question, we defined the questions into 2 groups, which were group of 2–3 terms per question and group of 4–6 terms per question. We can see that the exact type and the group of 2–3 terms performed better than the descriptive type and the group of 4–6 terms both in precision and recall.

4 Conclusions and Future Works

In this paper a proposed fuzzy TF-IDF weighting in synonym for Diabetes question and answers for compared the ability of the similarity of meaning by analysis similar question and answer. The system is able to choose answers from the corpus based on questions from the users. The theory of term weighting was applied through the fuzzy TF-IDF weighting method for the ranking of appropriate answers. To compare between the words in the sentences and in the answer documents, semantic similarity was utilized for retrieval of answer documents. Synonym words were used to solve the problems of different Thai sentences but similar meanings. Regarding the information for the efficiency evaluation of the system, thirty frequently asked questions and answers in diabetes care were collected to answer to meet user's needs. The most essential thing of this question answering systems is the understanding of user's questions with an objective to meet the user's needs. It could be seen that the evolution of research results was undertaken only by fuzzy TF-IDF weighing so that the results tended to provide the correct answers according to the algorithm. As a result, the best condition is Fuzzy degree = 1.0 of Top 1 whose precision is 92.50% and fuzzy degree = 0.8 and above as provided the accepted result as is 80.09%. The concept of this framework can be applied to questions and answers in a natural language. Other researchers can apply this framework to other question answering system similar to human language queries.

References

1. Bouziane, A., Bouchiha, D., Doumi, N., Malki, M.: Question answering systems: survey and trends. Proc. Comput. Sci. **73**, 366–375 (2015)
2. Braun, D., Hernandez-Mendez, A., Matthes, F., Langen, M.: Evaluating natural language understanding services for conversational question answering systems. In: Proceedings of the 18th Annual SIGdial Meeting on Discourse and Dialogue, pp. 174–185 (2017)
3. Mishra, A., Jain, S.K.: A survey on question answering systems with classification. J. King Saud Univ.-Comput. Inf. Sci. **28**(3), 345–361 (2016)
4. Irfan, M., Zulfikar, W.B.: Implementation of fuzzy C-means algorithm and TF-IDF on English journal summary. In: 2017 Second International Conference on Informatics and Computing (ICIC), pp. 1–5. IEEE, November 2017
5. Romero-Córdoba, R., Olivas, J.A., Romero, F.P., Alonso-Gonzalez, F., Serrano-Guerrero, J.: An application of fuzzy prototypes to the diagnosis and treatment of fuzzy diseases. Int. J. Intell. Syst. **32**(2), 194–210 (2017)
6. Falomir, Z., Olteţeanu, A.M.: Logics based on qualitative descriptors for scene understanding. Neurocomputing **161**, 3–16 (2015)
7. El-Sappagh, S., Elmogy, M., Riad, A.M.: A fuzzy-ontology-oriented case-based reasoning framework for semantic diabetes diagnosis. Artif. Intell. Med. **65**(3), 179–208 (2015)
8. Abacha, A.B., Zweigenbaum, P.: MEANS: a medical question-answering system combining NLP techniques and semantic Web technologies. Inf. Process. Manag. **51**(5), 570–594 (2015)
9. Terol, R.M., Martínez-Barco, P., Palomar, M.: A knowledge based method for the medical question answering problem. Comput. Biol. Med. **37**(10), 1511–1521 (2007)
10. Wu, Y., Wu, W., Li, Z., Zhou, M.: Mining query subtopics from questions in community question answering. In: AAAI, pp. 339–345, January 2015
11. Weis, K.H.: A case based reasoning approach for answer reranking in question answering. arXiv preprint arXiv:1503.02917 (2015)
12. Alodadi, M., Janeja, V.: Similarity in patient support forums using TF-IDF and cosine similarity metrics. In: 2015 International Conference on Healthcare Informatics (ICHI), pp. 521–522. IEEE (2015)
13. Roberts, D.W.: Ordination on the basis of fuzzy set theory. Vegetatio **66**(3), 123–131 (1986)
14. Salton, G.: Automatic processing of foreign language documents. J. Assoc. Inf. Sci. Technol. **21**, 187–194 (1970)
15. Breck, E., Burger, J.D., Ferro, L., Hirschman, L., House, D., Light, M., Mani, I.: How to evaluate your question answering system every day and still get real work done. arXiv preprint cs/0004008 (2000)
16. Garmendia, L.: The evolution of the concept of fuzzy measure. In: Intelligent Data Mining, pp. 185–200. Springer (2005)
17. Phetkrachang, K., Kittiphattanabawon, N.: Thai question answering systems in diabetes using logical co-operators. In: Proceedings of the 12th International Conference on Knowledge, Information and Creativity Support Systems, pp. 155–160. IEEE (2017)

Feature Comparison for Automatic Bug Report Classification

Bancha Luaphol[1(✉)], Boonchoo Srikudkao[1], Tontrakant Kachai[1],
Natthakit Srikanjanapert[1], Jantima Polpinij[1], and Poramin Bheganan[2(✉)]

[1] Intellect Laboratory, Department of Computer Science, Faculty of Informatics,
Mahasarakham University, Mahasarakham province, Thailand
`bancha.lu@ksu.ac.th,`
`{boonchoo.sri,tontrakant.kac,natthakit.sri,jantima.p}@msu.ac.th`
[2] Computer Science Program, Mahidol University International College,
Mahidol University, Nakhonpathom province, Thailand
`poramin.bhe@mahidol.ac.th`

Abstract. Nowadays, various bug tracking systems (BTS) such as Jira,
Trace, and Bugzilla have been developed and proposed to gather the
issues from users worldwide. This is because those issues, called bug
reports, contain a significant information for software quality maintenance and improvement. However, many bug reports with poor quality
might have been submitted to the BTS. In general, the reported bugs in
the BTS are firstly analyzed and filtered out by bug triagers. However,
with the increasing amount of bug reports in the BTS, manually classifying bug reports is a time-consuming task. To address this problem,
automatically distinguishing of bugs and non-bugs is necessary. To the
best of our knowledge, this task is never easy for bug reports classification because the problem of bug reports misclassification still occurs
to date. The background of this problem may be arise from using inappropriate or confusing features. Therefore, this work aims to study and
discover the most proper features for binary bug report classification.
This study compares seven features such as unigram, bigram, camel case,
unigram+bigram, unigram+camel case, bigram+ camel case, and all features together. The experimental results show that the unigram+camel
case should be the most proper features for binary bug report classification, especially when using with the logistic regression algorithm.
Consequently, the unigram+camel case should be the proper feature to
distinguish bug reports from the non-bugs ones.

Keywords: Bug report features · Bug report classification ·
Misclassification · Bug and non-bug · Unigram · Bigram · Camel case ·
Naïve Bayes · Logistic regression · Support vector machines

1 Introduction

Today, many bug tracking systems have been developed and proposed since it
should be a better way to collect the extensive bug reports and requirements

© Springer Nature Switzerland AG 2020
P. Boonyopakorn et al. (Eds.): IC2IT 2019, AISC 936, pp. 69–78, 2020.
https://doi.org/10.1007/978-3-030-19861-9_7

from numerous users worldwide [1–7]. As a result, systems like Bugzilla, Jira, Mantis or Trac are widely used for reporting software bugs [1–7]. However, many bug reports with poor quality (e.g. non-bug) are also submitted to the BTS. The poor quality reports will superfluously increasing time spent on the maintenance, development, and testing tasks. Therefore, bug triagers will screen and prioritize the bug reports before assigning suitable developers to fix a particular bug [2, 3,5,7]. Unfortunately, it can be estimated that more than 350 bug reports are submitted to the bug repository in the BTS every day, but the bug triagers are able to screen and prioritize only 300 bug reports per day [8]. This can demonstrate that classifying bug reports manually is a time consuming task. As the result, automatic bug report classification between bugs and non-bugs is necessary.

However, the previous studies have found that a main problem of automatic bug report classification misclassified between bugs and non-bugs. Also, this problem has occurred and many researchers are still studied to date [9–13]. Those researchers aimed to find better solutions for handling bug report misclassification. Based on the previous studies [11–13], 33.8% of all bug reports are confirmed that they can be misclassified [11]. The background of this problem may be arise from using of inappropriate or confusing features [11–13]. To the best of our knowledge, there are numbers of studies regarding bug report features for automatic bug report classification have been proposed, but those papers show promising results when using a standard dataset, called Herzig's dataset [11]. However, there has not been currently confirmed that, which bug report features are appropriate to be used with the real-world bug reports when distinguishing the bug reports from the non-bugs ones. Therefore, this is crucial and becomes a challenge for our study, where we aim to study and look for the required features of bug reports. The features such as unigram, bigram, camel case, combination between two of them (i.e. unigram+bigram, unigram+camel case, and unigram+camel case), and all of these features are considered in this study. In addition, we will study for two situations. The first study will experiment with the Herzig's dataset, while the second study will experiments with the real-world bug reports.

The rest of the paper is organized as follows. Section 2 describes the dataset, while Sect. 3 explains the research method. Section 4 presents the experimental results and discussion. Finally, Sect. 5 is the conclusion.

2   Dataset

There are two datasets used for this study. The first one is a standard dataset called Herzig's dataset [11]. The corpus contains 7,401 bug reports from five open sources. Two of them are from Bugzilla and the rest are from Jira. This dataset is formatted as csv. However, we select bug reports from the Herzig's dataset i.e. HttpClient, Lucene, and Jackrabbit for our study. The second dataset is downloaded bug reports relating to Mozilla Firefox (https://www.bugzilla.mozilla.org/) on October 1, 2017. We select only bug reports with the status of "new", "assigned", "reopen", "resolved", "verified", and "closed". The second dataset is also formatted as csv. Finally, the summary of our datasets used in this study can be shown as Table 1.

In general, a bug report contains two major parts i.e. summary and description. Then, the summary part actually is the title represented in the bug report, while the description part is the explanation of that particular bug report. However, to the best of our knowledge, many studies use only the summary part for their studies because this part contains less noise [2, 14–17]. Finally, we select 1,000 bug reports per class to be the training set and 300 bug reports per class to be the test set from each dataset.

Table 1. Summary of the datasets

Type of dataset	Example	Total of bug reports	Total of non-bug reports
Herzig's dataset	HttpClient	305	440
	Lucence	697	1,744
	Jackrabbit	937	1,464
Real-word dataset	Mozilla Firefox	2,000	2,000

3 Definition of Bug Report Features

In general, many works of bug report classification use unigram words as their bug report features. However, we study seven types of possible bug report features in this work in order to recognize the most appropriate feature type for the bug report classification. Those feature types are unigram, bigram, camel case, unigram+bigram, unigram+camel case, unigram+camel case, and the all features types i.e. unigram+bigram+camel case concurrently. Then, the unigram is a single word, and the bigram is a list of two consecutive words in a sentence. In this study, our bigram should be obtained from the noun group. Meanwhile, the camel case is a concatenation of words [18]. Some examples of the camel case are "ToolBar", "Menu-bar", "isCommitable", and so on. Then, these words are split into their components (or words) based on the occurrences of capital letter, hyphen symbol ("-"), underscore symbol ("_"), or period symbol ("."). The alternative names of camel case are "concatenation of words" [19] and "compound words" [20]. The examples of those feature types can be shown as Table 2.

4 Preliminary: Bigram Rule Modeling

This section describes the process of generating the bigram rules that will be used to find and extract all bigrams in a bug report. In this study, we deploy the concept of N-gram with Markov Models [21] to generate the bigram rules, where an N-gram is a subsequence of N items from a given sequence. The N-gram

Table 2. The examples of those bug report feature types

Feature types	Examples
Unigram	Fail, payload, method, incorrect
Bigram	Fields corrupts, thread hazard, different query
Camel case	LogSource.setLevel, XMLPersistenceManager

technique estimates the likelihood of w occurring in the new text. This estimation is based on the general frequency of the occurrence of w in the corpus. Therefore, this study applies the probability function (P), that is $P(w_n|w_1,\ldots,w_{n-1})$ to predict the next word-tag in form of bigram. The results of learning are called bigram rules that are used to extract the noun group. In this part, we utilize the Stanford Natural Language Inference corpus (SNLI) [22] to find the bigram rules of the noun group. This stage begins with POS tagging process. We also use the Stanford Parser [23] to extract all possible sequences of the POS tags of the noun group. In prior to learn the patterns of the noun-group by using Markov Models, these particular POS tag sequences need to augment each sequence of word tag first. A special symbol <S> must be applied at the beginning of the word-tag to give the N-gram context of the first word. In addition, a special end-symbol </S> is required at the end of that possible sequence. (See in Fig. 1).

Fig. 1. An illustration of possible word-tag sequences

The common formula for the bigram approximation to the conditional probability of the next word-tag (w) in a sequence can be shown as following:

$$P(w_n|w_1^{n-1}) \gg P(w_n|w_{n-N+1}^{n-1}) \tag{1}$$

The calculation of the probability for a complete word-tag sequence is illustrated in the following formula.

$$P(w_1^n) \approx \prod_{k=1}^{n} P(w_k|w_{k-1}) \tag{2}$$

In order to estimate the probability of word-tags co-occurrences from the corpus, the maximum likelihood estimation (MLE) could be applied to estimate the parameters of the bigram model. Subsequently, number of counts will be normalized so that the values will be between 0 and 1. An example of estimating a bigram value (or a certain bigram probability) of a word-tag, where w_{n-1} is the previous word-tag of a word-tag w_n. It will compute the count of the bigram $C(w_{n-1}, w_n)$. Then, it is normalized by the sum of all the bigrams that share the same first word-tag w_{n-1}. The formula can be:

$$P(w_n|w_{n-1}) = \frac{C(w_{n-1}w_n)}{\sum_w C(w_{n-1}w_n)} \qquad (3)$$

The examples of noun pattern are "NN NN", "JJR NN" and "JJ NN". These patterns are called bigram rules. After pruning bigram rules, we obtain 57 rules that are used for this study.

5   The Framework

This section describes the framework of our study. There are two main processing stages: bug report pre-processing, and bug report classifier modeling. Each stage can be detailed as follows.

5.1   Bug Report Pre-processing

To model bug report classifiers, it commences with the bug report pre-processing. Each bug report in the training set is divided into a meaning unit. In this study, the meaning unit of language can be word formatted i.e. unigram word, bigram words, camel case.

For segmenting bug report into *unigram words*, we utilize boundary markers such as the word spaces of written English. Meanwhile, we segment each bug report into *bigram words* using bigram rules generated from the preliminary stage. Then, each bug report commences by performing the POS tagging, and then the bigram rules are used to extract bigram words from the bug reports. If the bigram rules match with a noun group in the bug reports, that noun group will be extracted. For the *camel case* words, many term-words defined in a bug report are actually a concatenation of words. We also developed the Camel Case tokenizer to perform the tokenization for pre-processing the bug reports. To describe how to split the camel case word, consider the word "*ToolBar*" that is actually composed of two words. They are "*tool*" and "*bar*". Therefore, these composite term-words are split into two individual term-words.

Finally, those bug reports are represented as a vector space model, called bag of words (BOW). In addition, words in the BOW are weighted by a weighting scheme. This study applies only $tf(term frequency)$ weighting schemes, because this weighting scheme has been mentioned by [9, 10] that it is good enough for bug report classification. This is because certain terms such as "*failure*", "*crash*", or "*should*" are found in multiple documents. Then, those features can guide to classify differentiate bugs from non-bugs [9].

In this study, tf is the number of times that a term t occurs in a document d, denoted as $tf_{t,d}$. The formula of $tf_{t,d}$ can be calculated with logarithmically scaled frequency. It presents as follows.

$$tf_{t,d} = log(1 + tf_{t,d}) \qquad (4)$$

5.2   Bug Report Classifier Modeling

To create the bug report classifiers, suppose D is a training set of bug reports and d is a bug report, denoted as $D = d_1, d_2, \ldots, d_i$. A fixed set of classes can be denoted as $C = c_1, c_2, \ldots, c_i$. Therefore, a predicted class is $c \in C$. Then, C is the set of binary classes: real-bug and non-bug. In addition, there are three machine learning algorithms used for this study. Each algorithm can brief as follows.

Naïve Bayes (NB). It is a probabilistic classifiers based on applying Bayes' theorem with strong independence assumptions between the features [24]. Let v be classes (real-bug and non-bug), and w be the bug report features. The posterior probabilities $P(w|v) = P(w_1, w_2, w_3, \ldots, w_n|v)$ can be estimated directly from training data, but are generally infeasible to estimate unless the available data is vast. Therefore, the individual features are conditionally independent of each other given the classification can be:

$$P(w_1, w_2, ..., w_n|v) \cdot P(v|m) = \prod_{i=1}^{n} P(d_i|v) \tag{5}$$

According to the assumption above, the formula (5) becomes the Naïve Bayes Classifier that is the formal pattern used to distinguish of bugs and non-bugs. To predict the class of a new bug report, the following formula is used.

$$v_{NB} = \arg\max P(v) \prod_{i=1}^{n} P(d_i|v) \tag{6}$$

Support Vector Machine (SVM). It is a discriminative classification algorithm. This algorithm estimates the maximum margin around the separating hyperplane between classes [25]. The common concept of the SVM is to build a function that assigns the value $+1$ in a *"relevant"* region catching majority of the data points, while assigns -1 elsewhere. All hyperplanes (H) in $\Re^d$ are parameterized by a vector (w) and a constant b. The formula can be shown as:

$$w \cdot x + b = 0 \tag{7}$$

However, to find that specific hyperplane, the formula is:

$$f(x) = sign(w \cdot x + b) \tag{8}$$

In this study, the above formula is used to classify the bug reports correctly. It can define the hyperplane as follows. Suppose H_1 is $x_i \times w + b \geq +1$, when $y_i = +1$. Meanwhile, H_2 is $x_i \times w + b \leq 1$, when $y_i = -1$. If the bug reports are on H_1 and H_2, those bug reports are called the support vectors.

However, the margin (m) is used to separate the hyperplane. It is denoted as $d_x + d_y$, where d_x is the shortest margin to the closest positive point and d_y is the shortest margin to the closest negative point. Definitely, the margin between H_1 and H_2 can be defined as follows.

$$m = \frac{2}{\|w\|} \tag{9}$$

In order to maximize the margin, it can minimize $\|w\|$ based on the condition that, there are no data points (or bug report) between H_1 and H_2. To improve the quality of the SVM classifier, a kernel function is often used, where the kernel function enables them to operate in a high-dimension. Then, it can be implicit feature space without ever computing the coordinates of the data in that space. In this study, the radial basis function (RBF) is applied as the kernel function.

Logistic Regression (LR). It is a binary classifier that returns a probability score [26]. The basic formula of LR is:

$$P(c|D) = \frac{1}{1 + \exp^{(-z)}} \tag{10}$$

Consider the formula (10), let z be $(b_0 + w_1 b_1 + w_2 b_2 \cdots + w_i b_i)$ where b_i is used to normalize the weight of words i found in class c. Basically, the score used to normalize the weight of words i is called a coefficient score. This score is calculated from the estimation of Maximum Likelihood [26]. Commonly, the threshold used for the classification should be 0.5 [9,10,27]. Therefore, if $P(c|bug\ report) \geq 0.5$, this could be explained that the report is an actual-bug report while the other could be a non-bug report.

6   The Experimental Results and Discussion

After testing via the recall (R), precision (P), and F-measure (F1), the results of classifying differentiate bugs from non-bugs can be presented in Tables 3 and 4.

To the best of our knowledge, many previous studies that used the Herzig's dataset are often deployed the unigram terms as their features to distinguish bugs from non-bugs [9,10,14,17,28]. However, our experiments results that are represented in Table 3 reveal the little improved results than those of [10,13,29], where unigram+bigram and unigram+camel case are deployed in our experiments with a smaller dataset. Nevertheless, it can be found that the most appropriate algorithm is still the LR.

Definitely, certain previous studies also report the similar results from this algorithm [9,10,29]. In addition, consider the results of the classification based on the LR algorithm. It can be seen that other feature types except bigram return the satisfactory outcomes. The reason for unsatisfactory results in using bigram features may be caused by the inadequacy of the obtained features in the bigram form. Unsurprisingly, it is incomprehensible to distinguish bug reports and non-bug reports. Furthermore, it is possible that it cannot extract bigram

Table 3. The experimental results with Herzig's dataset

Bug report features	NB			LR			SVM (RBF)		
	R	P	F1	R	P	F1	R	P	F1
Unigram	**0.77**	**0.77**	**0.77**	**0.77**	**0.77**	**0.77**	**0.65**	**0.61**	**0.63**
Bigram	0.58	0.54	0.56	0.56	0.53	0.55	0.59	0.51	0.54
Camel case	0.76	0.75	0.76	**0.77**	**0.77**	**0.77**	0.62	0.60	0.61
Unigram+Bigram	**0.77**	**0.77**	**0.77**	**0.77**	**0.77**	**0.77**	0.61	0.52	0.56
Unigram+Camel case	**0.77**	**0.77**	**0.77**	**0.77**	**0.77**	**0.77**	0.62	0.58	0.60
Bigram+Camel case	0.76	0.76	0.76	**0.77**	**0.77**	**0.77**	0.59	0.52	0.56
All features	0.76	0.76	0.76	**0.77**	**0.77**	**0.77**	0.60	0.53	0.56

features from some bug reports because those bug reports does not match with the bigram rules. In this case, those bug reports will be removed from the corpus.

Consider Table 4. It can be seen that applying of our proposed features can be used for the real-world bug reports downloaded from the Mozilla Firefox without additional manual refinement, where all outputs return the satisfactory results. Especially, when these features are used in conjunction with the LR algorithm and the experimental results are close to those of [9, 10, 29] despite the data used in our study are smaller.

Table 4. The experimental results with the real-word bug reports of Mozilla Firefox

Bug report features	NB			LR			SVM (RBF)		
	R	P	F1	R	P	F1	R	P	F1
Unigram	0.56	0.66	0.61	0.65	0.67	0.66	0.62	0.66	0.64
Bigram	0.53	0.64	0.58	0.63	0.65	0.64	0.60	0.64	0.62
Camel case	0.53	0.63	0.58	0.63	0.65	0.64	0.61	0.63	0.62
Unigram+Bigram	0.62	0.67	0.64	0.71	0.71	0.71	0.67	0.67	0.67
Unigram+Camel case	**0.65**	**0.69**	**0.67**	**0.72**	**0.73**	**0.72**	**0.69**	**0.69**	**0.69**
Bigram+Camel case	0.60	0.64	0.62	0.69	0.69	0.69	0.67	0.67	0.67
All features	**0.65**	**0.69**	**0.67**	0.71	0.72	0.71	**0.69**	**0.69**	**0.69**

However, from both Tables 3 and 4, they can be seen that the deployed and efficient features used for these two datasets are unigram+camel case.

7 Conclusion

In the previous studies, there has not been currently confirmed that, which bug report features are appropriate to be used with the real-world bug reports when distinguishing bug reports from the non-bugs ones. This becomes a challenge for our study, where we aim to seek the desired features of bug reports. Then, features such as unigram, bigram, camel case, and combination between two of them (i.e.unigram+bigram, unigram+camel case, and unigram+camel case) are considered. Moreover, we will study for two situations. The first study will experiment with the Herzig's dataset, while the second study will experiment with the real-world bug report dataset. After experimenting by using two datasets, it can be shown that these features in conjunction with the LR algorithm return the satisfactory outputs. However, we found that the unigram+camel case should be the most appropriate features for both datasets. Therefore, the unigram+camel case should be the suitable feature to distinguish bug reports from the non-bugs ones.

References

1. Sandusky, R.J., Gasser, L., Ripoche, G.: Bug report networks: varieties, strategies, and impacts in a F/OSS development community. In: The 1st International Workshop on Mining Software Repositories, pp. 80–84 (2004)
2. Jalbert, N., Weimer, W.: Automated duplicate detection for bug tracking systems. In: IEEE International Conference on Dependable Systems and Networks With FTCS and DCC, pp. 52–61 (2008)
3. Wang, X., Zhang, L., Xie, T., Anvik, J., Sun, J.: An approach to detecting duplicate bug reports using natural language and execution information. In: ACM/IEEE 30th International Conference on Software Engineering, pp. 461–470 (2008)
4. Bhattacharya, P., Neamtiu, I.: Bug-fix time prediction models: can we do better? In: Proceedings of the 8th Working Conference on Mining Software Repositories, pp. 207–210 (2011)
5. Tian, Y., Sun, C., Lo, D.: Improved duplicate bug report identification. In: The 16th European Conference on Software Maintenance and Reengineering (CSMR), pp. 385–390 (2012)
6. Zhang, J., Wang, X., Hao, D., Xie, B., Zhang, L., Mei, H.: A survey on bug-report analysis. Sci. China Inf. Sci. **58**(2), 1–24 (2015)
7. Aggarwal, K., Timbers, F., Rutgers, T., Hindle, A., Stroulia, E., Greiner, R.: Detecting duplicate bug reports with software engineering domain knowledge. J. Softw.: Evol. Process. **29**(3), e1821 (2017)
8. Anvik, J., Murphy, G.C.: Reducing the effort of bug report triage: recommenders for development oriented decisions. ACM Trans. Softw. Eng. Methodol. **20**, 10:1–10:35 (2011)
9. Antoniol, G., Ayari, K., Di Penta, M., Khomh, F., Guéhéneuc, Y.-G.: Is it a bug or an enhancement?: A text-based approach to classify change requests. In: Proceedings of the 2008 Conference of the Center for Advanced Studies on Collaborative Research: Meeting Of Minds. ACM (2008)
10. Pingclasai, N., Hata, H., Matsumoto, K.-I.: Classifying bug reports to bugs and other requests using topic modeling. In: The 20th Asia-Pacific Software Engineering Conference (APSEC), pp. 13–18. IEEE (2013)
11. Herzig, K., Just, S., Zeller, A.: It's not a bug, it's a feature: how misclassification impacts bug prediction. In: The 35th International Conference on Software Engineering (ICSE), pp. 103–104 (2013)

12. Limsettho, N., Hata, H., Monden, A., Matsumoto, K.: Automatic unsupervised bug report categorization. In: The 6th International Workshop on Empirical Software Engineering in Practice (IWESEP), pp. 7–12. IEEE (2014)

13. Qin, H., Sun, X.: Classifying bug reports into bugs and non-bugs using LSTM. In: Proceedings of the Tenth Asia-Pacific Symposium on Internetware. ACM (2018)

14. Zhou, Y., Tong, Y., Gu, R., Gall, H.: Combining text mining and data mining for bug report classification. In: 2014 IEEE International Conference on Software Maintenance and Evolution, pp. 311–320. IEEE (2014)

15. Lamkanfi, A., Demeyer, S., Giger, E., Goethals, B.: Predicting the severity of a reported bug. In: 2010 7th IEEE Working Conference on Mining Software Repositories (MSR 2010), pp. 1–10. IEEE (2010)

16. Ko, A.J., Myers, B.A., Chau, D.H.: A linguistic analysis of how people describe software problems. In: Visual Languages and Human Centric Computing (VL/HCC 2006), pp. 127–134 (2006)

17. Pandey, N., Hudait, A., Sanyal, D.K., Sen, A.: Automated classification of issue reports from a software issue tracker. Presented at the Progress in Intelligent Computing Techniques: Theory, Practice, and Applications (2018)

18. Almhana, R., Mkaouer, W., Kessentini, M., Ouni, A.: Recommending relevant classes for bug reports using multi-objective search. In: Proceedings of the 31st IEEE/ACM International Conference on Automated Software Engineering, pp. 286–295 (2016)

19. Zhou, J., Zhang, H., Lo, D.: Where should the bugs be fixed? More accurate information retrieval based bug localization based on bug reports. In: Proceedings of the 34th International Conference on Software Engineering, pp. 14–24 (2012)

20. Ye, X., Bunescu, R., Liu, C.: Mapping bug reports to relevant files: a ranking model, a fine-grained benchmark, and feature evaluation. IEEE Trans. Softw. Eng. **42**(4), 379–402 (2016)

21. Indurkhya, N., Damerau, F.J.: Handbook of Natural Language Processing. CRC Press, New York (2010)

22. Bowman, S.R., Angeli, G., Potts, C., Manning, C.D.: A large annotated corpus for learning natural language inference. https://nlp.stanford.edu/pubs/snli_paper.pdf

23. De Marneffe, M.C., MacCartney, B., Manning, C.D.: Generating typed dependency parses from phrase structure parses. In: Proceedings of LREC, pp. 449–54 (2006)

24. Nizamani, Z.A., Liu, H., Chen, D.M., Niu, Z.: Automatic approval prediction for software enhancement requests. Autom. Softw. Eng. **25**, 347–381 (2017). https://doi.org/10.1007/s10515-017-0229-y

25. Mitra, V., Wang, C.-J., Banerjee, S.: Text classification: a least square support vector machine approach. Appl. Soft Comput. **7**, 908–914 (2007)

26. Webb, AR., Copsey, K.D.: Statistical Pattern Recognition, 3rd edn (2011). https://doi.org/10.1002/9781119952954

27. Pandey, N., Sanyal, D.K., Hudait, A., Sen, A.: Automated classification of software issue reports using machine learning techniques: an empirical study. Innov. Syst. Softw. Eng. **13**, 1–19 (2017)

28. Du, X., Zheng, Z., Xiao, G., Yin, B.: The automatic classification of fault trigger based bug report. In: 2017 IEEE International Symposium on Software Reliability Engineering Workshops (ISSREW), pp. 259–265. IEEE (2017)

29. Terdchanakul, P., Hata, H., Phannachitta, P., Matsumoto, K.: Bug or not? Bug report classification using N-gram IDF. In: Proceedings 2017 IEEE International Conference on Software Maintenance and Evolution, ICSME 2017, pp. 534–538. IEEE (2017)

A Novel Three Phase Approach for Single Sample Ear Recognition

Nitin Kumar[(⊠)]

Department of Computer Science and Engineering,
National Institute of Technology, Srinagar, Uttarakhand, India
nitin@nituk.ac.in

Abstract. Ear recognition becomes challenging when only single training sample is available. In such scenarios, most of the existing ear recognition methods fail to work because of insufficient training data. In this
paper, we propose a novel three phase approach for single sample ear
recognition where in phase 1, ear images are normalized using histogram
equalization, in phase 2, a novel ear representation called Two Dimensional (2D) Eigenears is proposed and in the last phase, classification
is carried out using nearest neighbour classifier. The proposed approach
is the first approach in single sample ear recognition which is based
only on two dimensional (2D) ear images. Here, normalized ear image
of each identity is represented as a linear combination of the so called
2D Eigenears thereby reducing the time complexity of the proposed approach. Experimental results on two publicly available ear datasets viz.
IIT Delhi and CP show the efficacy of the proposed approach. The proposed method achieves more than 70% and approximately 80% rank-1
recognition accuracy on IIT Delhi and CP databases respectively.

Keywords: Biometric · Small sample size · Linear · Eigenvectors ·
Phase

1 Introduction

Biometrics [1] such as iris, face, fingerprint, voice, gesture, gait etc. are being
increasingly used for person identification. Ear [2] is a recent biometric and possess certain advantages over other biometrics such as invariability of ear with
time, non-cooperation of the users, uniqueness characteristic across individuals
and so on. Identity recognition using ear image is an active research area. The
pioneering work in the field of ear recognition was presented by [3] in which he
had studied 10,000 ears of different people and it was observed that none of them
matched with any other. Since then, several methods have been suggested in literature for ear recognition. A comprehensive survey on ear recognition has been
carried out by [2] and they broadly divided ear recognition methods into four categories viz. (i) Geometric methods (ii) Holistic methods (iii) Local methods and
(iv)Hybrid methods. Geometric methods consists of earlier methods suggested

© Springer Nature Switzerland AG 2020
P. Boonyopakorn et al. (Eds.): IC2IT 2019, AISC 936, pp. 79–88, 2020.
https://doi.org/10.1007/978-3-030-19861-9_8

for ear recognition such as [4,5] were geometric methods in which edge detection was employed to find the edges in the ear image and then the distance between these features was used for recognition. Geometric methods tend to be simple but due to dependency on edge detection, suffer from illumination and noise. Holistic methods are based on feature extraction from gray level ear images. Some of the popular methods under this category are Principal Component Analysis (PCA) suggested by [6], Independent Component Analysis (ICA) proposed by [7], Full Space Linear Discriminant Analysis given by [8] and Sparse representation [9]. Holistic methods possess comparable performance if the ear images are properly aligned and pre-processed, otherwise the recognition accuracy of these methods is negatively affected in general. Local methods such as Scale-Invariant Feature Transform (SIFT) suggested by [10], Speedup Robust Features (SURF) proposed by [11] extract features from the local regions or patches in an ear image. These local features may represent sparse or dense encoding of local regions. These methods give better result even if the images are not properly aligned. However, the computational complexity of these methods is more than holistic methods. Hybrid methods suggested by Morales et al. [12] and Benzaoui et al. [13] combine advantages of one or more methods in other categories. However, these methods are more complex than holistic or local methods. All the ear recognition methods discussed above require more than one ear image per person for training. Methods such as [8] fail altogether when only single training image is available. Other methods such as proposed by [6,7] require more than one training images for better performance. Despite these issues, single sample ear recognition also possesses some advantages such as less storage and computational requirements.

1.1 Contribution

Single sample ear recognition has received much less attention from the research community. To the best of our knowledge, there is single work reported in literature by Chen et al. [14] which addresses single sample ear recognition problem. Moreover, their method employs both 2D and 3D ear information for identity recognition. There is no method in literature which solves single sample ear recognition problem using only 2D images. In this paper, we address single sample ear recognition problem by suggesting a novel three phase approach for single sample ear recognition. In phase 1, ear images are normalized using histogram equalization. In phase 2, a novel representation called 2D Eigenears is derived from the available single training samples and each ear image is represented as a linear combination of these 2D Eigen ears. In last phase, recognition is performed using nearest neighbour classifier. By employing the proposed representation, features are directly extracted from the 2D ear images of the person thereby greatly reducing the time complexity. The rest of the contents in the manuscript is organized as follows: Sect. 2 provides a brief overview of Eigen ear representation suggested in literature. In Sect. 3, mathematical formulation of the proposed method is described. The datasets along with the experimental results are presented in Sect. 4. Important conclusions are drawn based on experimental results and are given at the end in Sect. 5. Some future research directions are also given in Sect. 5.

2   Related Work

Let $\mathbf{x} \in \mathbb{R}^d$ be a column vector representing an ear image in a d dimensional space and there are c identities $\{1, 2, 3, ..., c\}$ such that each identity has N_i images. Thus, total number of training images is $N = \sum_{i=1}^{c} N_i$. Let $\bar{\mathbf{x}} = \frac{1}{N} \sum_{i=1}^{N} \mathbf{x}_i$ represents the mean ear vector of the training data. Hence, the total scatter matrix for the training data is given below [15]:

$$\mathbf{C} = \sum_{i=1}^{N} (\mathbf{x}_i - \bar{\mathbf{x}})(\mathbf{x}_i - \bar{\mathbf{x}})^T \tag{1}$$

[17] introduced the concept of Eigen ears in their research work. These eigen ears are the basis vectors representing the directions along which the variance in the ear images is maximum. The criterion function of this approach matches with that of Principal Component Analysis (PCA) [16] and is given by [15]:

$$J(\mathbf{W}_{pca}) = \overset{arg\ max}{\mathbf{W}_{pca}} \mathbf{W}^T \mathbf{C} \mathbf{W} \tag{2}$$

The solution to above criterion is given by eigenvalue decomposition of the total scatter matrix as follows:

$$\mathbf{CE} = \mathbf{E} \Lambda \tag{3}$$

where $\mathbf{E} = [\mathbf{e}_1 \quad \mathbf{e}_2 \quad ... \quad \mathbf{e}_d]$ is the eigenvector matrix with corresponding eigenvalue given by the diagonal matrix $\Lambda = diag(\lambda_1, \quad \lambda_2, \quad ..., \quad \lambda_d)$. The transformation matrix, $\mathbf{W}_{pca}$ is constructed by selecting k eigenvectors corresponding to the largest k eigenvalues and the training data samples are projected using W_{pca} as follows:

$$\mathbf{Y} = \mathbf{W}_{pca}^T \mathbf{X} \tag{4}$$

The test sample is also transformed using $\mathbf{W}_{pca}$ and recognition is performed. Afterwards, [6] presented a comparative study between Eigen faces and Eigen ears and found that there is not much difference between performance of these two representations. However, in the methods suggested by Chang et al. [6] and Victor et al. [17], it is required to convert the biometric image into one-dimensional column vector. This renders the size of total scatter matrix to be extremely large and hence requires huge computational resources. Moreover, the structural information is also not captured by Eigen ears. To overcome these limitations, Yang et al. [18] have suggested Two Dimensional Principal Component Analysis (2DPCA) in which the image matrices are directly employed for feature extraction without converting to column vectors.

3   Proposed Technique

In this section, we describe various phases involved in the proposed approach for single sample ear recognition. The proposed approach consists of three phases *viz.* (i) Normalization (ii) 2D Eigen Ear Learning (iii) Classification. A schematic representation of the proposed approach is shown in Fig. 1.

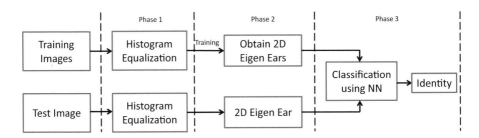

Fig. 1. Schematic representation of the proposed approach (NN - Nearest Neighbour)

3.1   Normalization

In the proposed approach, image normalization is carried out using Histogram Equalization [19]. The main objective in histogram equalization is to flatten the histogram of an image. Given an image $\mathbf{A}$ of size $a \times b$ in the gray scale with L levels. Suppose p_i is the probability of occurrence of i-th pixel intensity and is given by:

$$p_i = \frac{\sum_{i=1}^{a} \sum_{j=1}^{b} \delta(A_{ij}, i)}{a \times b} \tag{5}$$

where $\delta(x, y)$ is the kronecker delta function which is 1 is x and y are equal ans 0 otherwise.

The histogram equalized image $\mathbf{B}_{ij}$; $(i = 1, 2, ..., a; \; j = 1, 2, ..., b)$ is given by:

$$\mathbf{B}_{ij} = floor((L-1) \sum_{i=0}^{\mathbf{A}_{ij}} p_i) \tag{6}$$

The above equation is same as transforming the pixel intensities k of $\mathbf{A}$ by the function:

$$T(k) = floor((L-1) \sum_{i=0}^{k} p_i) \tag{7}$$

such that the transformed pixels are uniformly distributed on $[0, L-1]$ and thereby equalizing the histogram of the image. A sample ear image and its histogram equalized image is shown in Fig. 2. It can be readily observed that Histogram Equalization improves the appearance of the ear images.

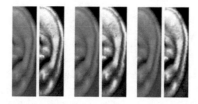

Fig. 2. Sample ear images and their corresponding Histogram equalized images of a person from IIT Delhi ear database

3.2   2D Eigen Ears

Here, we introduce a novel representation called Two Dimensional (2D) Eigen Ears for single sample ear recognition. In extreme small sample size situations where only single training sample per person is available, the performance of the methods suggested in [6] and [17] are severely affected as we will see in next section. There are several reasons for such a drop in performance:

(i) Low rank of the scatter matrix
(ii) Structural information of data samples is not performed
(iii) Poor estimation of the parameters of the scatter matrix.

Suppose $\mathbf{A}$ (size $= a \times b$) represents a two dimensional ear image of an identity and there are N identities in consideration. Let $\bar{\mathbf{A}}$ denotes the mean ear image among all the identities as given below:

$$\bar{\mathbf{A}} = \frac{1}{N} \sum_{i=1}^{N} \mathbf{A}_i \tag{8}$$

Now, the *image scatter matrix* of the training samples can be defined as:

$$\mathbf{S} = \sum_{i=1}^{N} (\mathbf{A}_i - \bar{\mathbf{A}})^T (\mathbf{A}_i - \bar{\mathbf{A}}) \tag{9}$$

then the criterion for maximizing the scatter in the transformed space is given below:

$$J(\mathbf{W}) = {}^{arg}_{\mathbf{W}} {}^{max} \mathbf{W}^T \mathbf{S} \mathbf{W} \tag{10}$$

This criterion matches with that of [18] with a difference that we have considered *image scatter matrix* intsead of *image covariance matrix* for formulating the criterion function. The optimal solution to criterion given in (10) is obtained by eigenvalue decomposition of $\mathbf{S}$. The projection matrix $\mathbf{W}$ is found by choosing k eigenvectors corresponding to top k eigenvalues *i.e.*

$$\mathbf{W} = [\mathbf{f}_1 \quad \mathbf{f}_2 \quad \dots \quad \mathbf{f}_k]; \quad k = 1, 2, \dots, a \tag{11}$$

Thus, the original image $\mathbf{A}$ is projected onto k eigenvectors found in the above equation as follows:

$$\mathbf{Y} = \mathbf{AW} \qquad (12)$$

The projected image $\mathbf{Y}$ has k column vectors known as principal components of the ear image $\mathbf{A}$. It is worth noting that each principal component in the proposed representation is a vector in contrast to Eigenears where principal component is a scalar.

3.3 Classification

Once the principal components are obtained by using the transformation (11), all the training samples and the test sample are transformed using $\mathbf{W}$. Afterwards, classification is performed using nearest neighbour rule. The distance between any two feature matrices $\mathbf{A}$ and $\mathbf{B}$ of size $b \times k$ obtained using the proposed technique is:

$$d(\mathbf{A}, \mathbf{B}) = \sum_{i=1}^{k} ||\mathbf{A}_i - \mathbf{B}_i||_2 \qquad (13)$$

where $\mathbf{A}_i$ and $\mathbf{B}_i$ are the i-th column of the feature matrices $\mathbf{A}$ and $\mathbf{B}$ respectively.Further, $||(.)||_2$ is the Euclidean norm between the feature vectors between $\mathbf{A}_i$ and $\mathbf{B}_i$.

Suppose that the feature matrices corresponding to training samples are represented as $\mathbf{A}_1, \mathbf{A}_2, ..., \mathbf{A}_c$ and the feature matrix corresponding to test image is $\mathbf{T}$. Then the minimum distance between the test sample and the training images is given by:

$$d(\mathbf{T}, \mathbf{A}_j) = {}^{arg\ min}_{i} d(\mathbf{T}, \mathbf{A}_i) \qquad (14)$$

Hence, the test sample is assigned to the class j to which the distance of the test sample is minimum.

3.4 Complexity Analysis

As evident from the mathematical formulation of the proposed method, 2D eigen ears are directly obtained from the ear images without being converted into 1D column vectors. This greatly reduces the size of scatter matrix.

The methods proposed by [6] and [17] involve complexity of $O(d^3)$ due to the fact that the covariance matrix is of size $d \times d$. In the proposed method, 2D Eigen ears are obtained using the scatter matrix of size $b \times b$. It is worth noting that $d = a \times b$ and hence the proposed method is having time complexity of $O(b^3)$ which is $O(a^3)$ times faster than the time complexity than the eigen ear method. This is a very important advantage of the proposed approach.

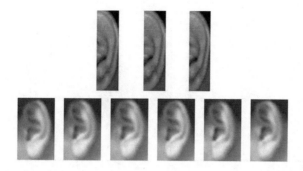

Fig. 3. Sample images from IIT Delhi (top row) and CP (bottom row) ear databases

4   Experimental Setup and Results

4.1   Databases Used

To demonstrate the efficacy of the proposed method, experiments have been carried out on IIT Delhi [20] and CP [21] Ear databases. IIT Delhi database ear consists of 375 cropped images of 125 identities with 3 images per identity. Each image in the cropped database is of size 180 × 50. CP [21] ear database consists of 102 images of 17 people with 6 images per person. In this database, the images have varying size. To make them suitable for experiments, all the images are resized to 62 × 30 which is the average image size of all the images in the database. Some sample images from both the databases are shown in Fig. 3.

4.2   Classification Accuracy

In the proposed approach, we need to choose the number of principal components (k). For this, we analyzed the eigenvalue spectrum of the proposed 2D eigen ears representation as shown in Fig. 4 for IIT Delhi and CP Ear databases. As we know that most of the variance is contained on the top few eigenvalues, we selected top 10 eigenvalues for forming the transformation matrix. These eigenvalues explain more than 99% and 98% variance on IIT Delhi and CP ear databases respectively.

For IIT Delhi ear database, only a single image of each identity is used as training and others are used for testing. As there are only 3 images per person, we have repeated this process three times and average classification accuracy is found using the proposed and compared methods. We have also varied the rank or the number of principal components to see the variation of recognition accuracy w.r.t the rank. The average classification accuracy is shown in Fig. 5. It can be readily observed that the proposed method gives better performance than the methods suggested by [6] and [7]. On CP ear database also, only a single image of each identity is used as training and others are used for testing. As there are 6 images per person, we have repeated this process six times and average classification accuracy is noted. The average classification accuracy is

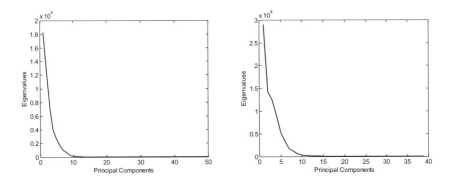

Fig. 4. Eigenvalue spectrum on IIT Delhi (left) and CP (right) ear database

shown in Fig. 5. On CP database also, the proposed method outperforms other methods.

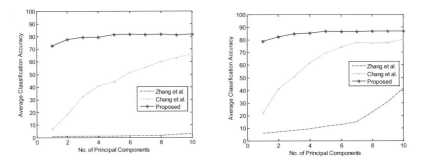

Fig. 5. Average classification accuracy on IIT Delhi (left) and CP (right) ear database

It is encouraging that rank-1 recognition of the proposed method is more than 70% on IIT Delhi database and approaching 80% on CP Ear database with only a single training image. This accuracy further approaches to 80% and 90% on IIT Delhi and CP databases respectively if rank is increased to 10. The performance of Eigen Ears is less in comparison to the proposed method in all cases.

4.3   Training Time

The average training time required by the proposed and compared algorithm is given in Table 1. It can be easily observed that the proposed method requires less training time than the methods suggested by Victor et al. [17] and Zhang et al. [7] on both the datasets. The method by Zhang et al. [7] takes maximum training time as there is no closer form solution to the criterion.

Table 1. Average training time (sec.) required by various methods

Methods/datasets	IIT Delhi	CP
Zhang et al. [7]	5.293	0.4575
Chang et al. [6]	0.0665	0.0030
Proposed	**0.0166**	**0.0023**

5 Conclusion and Future Work

In this paper, we proposed a novel three phase approach for single sample ear recognition. We also introduced a novel representation called 2D Eigenears which are more discriminative than the earlier Eigen ears suggested in literature specially in single sample scenarios. The discrimination is better due to the fact that each individual column of the face images is treated as a feature vector in contrast to the feature vector consisting of the whole image in 1-D methods. The classification is carried out using nearest neighbour rule and the time complexity of the proposed algorithm is much better than the methods suggested by [17] and [6]. The experimental results show that the proposed methods significantly outperform other methods on IIT Delhi and CP Ear databases. The proposed method achieves more than 70% and 80% rank-1 recognition accuracy on IIT Delhi and CP datasets. In Future, we shall propose novel feature extraction methods which can give better rank 1 recognition accuracy in single sample scenarios.

References

1. Jain, A.K., Ross, A., Prabhakar, S.: An introduction to biometric recognition. IEEE Trans. Circuits Syst. Video Technol. **14**(1), 4–20 (2004)
2. Emersic, Z., Struc, V., Peer, P.: Ear recognition: more than a survey. Neurocomputing **255**(Suppl. C), 26–39 (2017). Bioinspired intelligence for machine learning
3. Iannarelli, A.V.: Ear Identification. Forensic Identification Series. Paramont Publishing Company (1989)
4. Burge, M., Burger, W.: Ear biometrics, pp. 273–285. Springer, Boston (1996)
5. Moreno, B., Sanchez, A., Velez, J.F.: On the use of outer ear images for personal identification in security applications. In: Proceedings IEEE 33rd Annual 1999 International Carnahan Conference on Security Technology (Cat. No. 99CH36303), pp. 469–476 (1999)
6. Chang, K., Bowyer, K.W., Sarkar, S., Victor, B.: Comparison and combination of ear and face images in appearance-based biometrics. IEEE Trans. Pattern Anal. Mach. Intell. **25**(9), 1160–1165 (2003)
7. Zhang, H.J., Mu, Z.C., Qu, W., Liu, L.M., Zhang, C.Y.: A novel approach for ear recognition based on ICA and RBF network. In: 2005 International Conference on Machine Learning and Cybernetics, vol. 7, pp. 4511–4515 (2005)
8. Yuan, L., Mu, Z.: Ear recognition based on 2D images. In 2007 First IEEE International Conference on Biometrics: Theory, Applications, and Systems, pp. 1 – 5 (2007)

9. Zhang, B., Mu, Z., Li, C., Zeng, H.: Robust classification for occluded ear via Gabor scale feature-based non-negative sparse representation. Opt. Eng. **53**, 061702 (2013)
10. Dewi, K., Yahagi, T.: Ear photo recognition using scale invariant keypoints. In: Computational Intelligence (2006)
11. Prakash, S., Gupta, P.: An efficient ear recognition technique invariant to illumination and pose. Telecommun. Syst. **52**(3), 1435–1448 (2013)
12. Morales, A., Diaz, M., Llinas-Sanchez, G., Ferrer, M.A.: Earprint recognition based on an ensemble of global and local features. In: 2015 International Carnahan Conference on Security Technology (ICCST), pp. 253 – 258 (2015)
13. Benzaoui, A, Hezil, N., Boukrouche, A.: Identity recognition based on the external shape of the human ear. In: 2015 International Conference on Applied Research in Computer Science and Engineering (ICAR), pp. 1–5 (2015)
14. Chen, L., Mu, Z., Zhang, B., Zhang, Y.: Ear recognition from one sample per person. PLoS ONE **10**(5), 1–16 (2004)
15. Belhumeur, P.N., Hespanha, J.P., Kriegman, D.J.: Eigenfaces vs. Fisherfaces: recognition using class specific linear projection. IEEE Trans. Pattern Anal. Mach. Intell. **19**(7), 711–720 (1997)
16. Turk, M., Pentland, A.: Eigenfaces for recognition. J. Cogn. Neurosci. **3**(1), 71–86 (1991)
17. Victor, B., Bowyer, K., Sarkar, S.: An evaluation of face and ear biometrics. In: Object Recognition Supported by User Interaction for Service Robots, vol. 1, pp. 429–432 (2002)
18. Yang, J., Zhang, D., Frangi, A.F., Yang, J.: Two-dimensional PCA: a new approach to appearance-based face representation and recognition. IEEE Trans. Pattern Anal. Mach. Intell. **26**(1), 131–137 (2004)
19. Gonzalez, R.C., Woods, R.E.: Digital Image Processing, 3rd edn. Prentice-Hall Inc., Upper Saddle River (2006)
20. Kumar, A., Wu, C.: Automated human identification using ear imaging. Pattern Recogn. **45**(3), 956–968 (2012)
21. Carreira-Perpinan, M. A.: Compression neural networks for feature extraction: application to human recognition from ear images. M.Sc. thesis, Faculty of Informatics, Technical University of Madrid, Spain (1995). (in Spanish)

Comparison of Thai Sentence Sentiment Tagging Methods Using Thai Sentiment Resource

Kanlaya Thong-iad and Ponrudee Netisopakul[(✉)]

Faculty of Information Technology,
King Mongkut's Institute of Technology Ladkrabang, Bangkok, Thailand
Kanlaya.satin@gmail.com, ponrudee@it.kmitl.ac.th

Abstract. Opinion mining aims to determine text sentiment from social network. Thai sentiment resource [3] can be utilized to tag sentiment of Thai text. Thai sentiment resource consists of Thai terms, for each term a number of English synsets and a set of sentic values and a polarity. However, most terms have many synsets. In order to decide on a sentence emotion, six methods for Thai sentence sentiment tagging were proposed and evaluated. The dataset was obtained from social media comments of an airline service. For the six methods, tagging accuracies ranged from 64.7 to 73.6%. The best method used only adverb and adjective synsets to determine a sentence emotion; while the worst method used only verb synsets.

Keywords: Thai social comments · Sentence sentiment tagging ·
Thai sentiment resource

1 Introduction

Sentiment analysis is a task to determine whether a sentence expresses a positive or negative emotion: for example, it can analyze customer feedback or comment on social network or other social media sources. Currently, there are various English sentiment resources for tagging emotion [1].

We focus here on Thai sentiment analysis from Thai public sources. It is challenging because the Thai language is complex in structure of word and form of sentence, which results in ambiguity and limited resources. Currently, Thai public resource includes Sentiment Text Tagging System (STTS) [2] and Thai Sentiment Resource [3]. Although STTS can be used to tag Thai emotions, there are limits because normally one Thai word can be translated to many English words, each with different sentiment value. For example, the Thai word, "ร่ำรวย (Rårwy)", requires several English words, "rich" and "affluent", to capture its meaning. However, the Thai Sentiment Resource can overcome this limitation: it was created using Thai WordNet [4] and SenticNet4 [5]. Some single Thai words in this resource have more than one sentiment value, reflecting actual usage.

We used the Thai Sentiment Resource for finding sentiment values and evaluated six methods for tagging emotions from a social media dataset which are assigned to positive or negative emotions. Our work can be compared to Lertsuksakda *et al.*'s previous work [6].

© Springer Nature Switzerland AG 2020
P. Boonyopakorn et al. (Eds.): IC2IT 2019, AISC 936, pp. 89–98, 2020.
https://doi.org/10.1007/978-3-030-19861-9_9

This paper is organized as follows: Sect. 2 reviews the theory of Thai WordNet, SenticNet, Thai Sentiment Resource and Deep Cut. Section 3 describes the dataset for tagging emotions. Section 4 details the six methods used. Section 5 shows implementation results and Sect. 6 concludes and suggests possible future work.

2 Literature Review

2.1 Thai WordNet

Thai WordNet [4] was developed by Asian WordNet framework base on PWN (Princeton's WordNet): it uses English – Thai dictionaries for automatically relating English to Thai words. It created a word network website for use in other work.

2.2 SenticNet4

SenticNet is resource for opinion mining [1]: it collects opinion from text or sentences for classification as positive, negative or neutral. It was created by artificial intelligence (AI) and semantic web for finding an opinion's emotion and processing of natural language text at a semantic level. The current SenticNet4 [5] has been developed from previous versions created from ConceptNet, from data on the internet collected by volunteers in the Open Mind Common Sense project and kept in the Open Mind corpus [7]. SenticNet4 expands existing nouns and verbs on commonsense using generalization. This process made automatically hierarchical clustering and dimensional reduction for finding original nouns and verbs. The result is AffectiveSpace bring to assign semantic and sentiment values before collect to SenticNet4.

2.3 Thai Sentiment Resource

Thai Sentiment Resource [3] is our Thai resource created using Thai WordNet as the main resource. It has two functions: sentiment values assignment and new words formation.

The first step for assigning sentiment values selects Thai words from two dictionaries - Lexitron [8] and Volubilis [9]: words were checked for redundancy from both sources before addition to the database. Each Thai word corresponds to English synsets (synonym set) in Thai WordNet. One Thai word can have more than one English synset. Each English synset corresponds to a set of four sentic values and one polarity from SenticNet4. The four sentic values are four dimensions consisting of pleasantness, attention, sensitivity and aptitude. Thus sentiment values kept on the Thai Sentiment Resource database form a 5-tuple (pleasantness, attention, sensitivity, aptitude, and polarity).

The second component creates new words from original Thai words. The words from the Stop Word function were checked with the original Thai word. For example, the Thai word "ความสุข (Khwām sūkh)" (happy) may be treated as a compound word. If part of the original Thai word was in Stop Word set, remove this word from the original Thai word. In this case, the Stop Word is "ความ (Khwām)" and is removed leaving the new word "สุข (sūkh)" (which also means "happy") - the original Thai word with the Stop Word removed. The new word can be found in the English synsets and sentiment values are assigned using the same method as the original word described in the previous step.

The result is that one Thai word may have more than one synset. It reflects the real usage, in that one Thai word can have more than one English translation.

2.4 Deep Cut

Deep Cut is public tool for splitting a sentence into words: it is a Thai word tokenization library based on a deep neural network [10].

3 Dataset

We collected a dataset from airline comments on Facebook containing 1,153 comments, with manually tagged emotions. 401 comments were tagged as positive emotions versus 752 tagged as negative.

4 Methodology

We developed a scheme for automatically tagging emotions from Thai sentence shown in Fig. 1: it has two main steps:

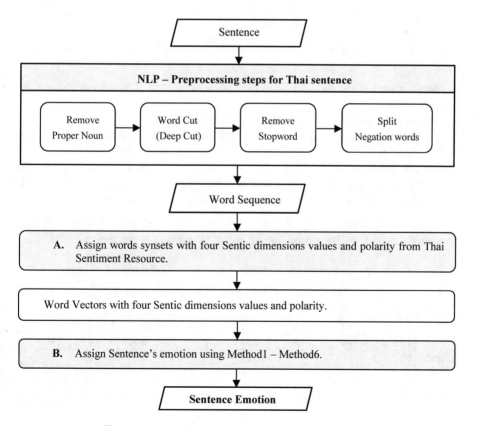

Fig. 1. Process flow for automatically tagging emotions

4.1 NLP-Preprocessing Steps for Thai Sentence

One sentence from the dataset is used as an example of the calculations for each of the six methods. The original Thai sentence is followed by a direct literal translation of it (Table 1).

Thai: ABC บริการ แย่มาก ยกเลิก เทียวบิน ไม่ แจ้ง ล่วงหน้า ค่ะ

English: ABC Service bad cancellation flight not inform ahead sir

Table 1. Sentence processing steps.

Steps	Result						
Proper noun	~~ABC~~ บริการแย่มาก ยกเลิกเทียวบินไม่แจ้งล่วงหน้าค่ะ						
removal	~~ABC~~ bad service cancellation flight not ahead sir						
Deep Cut	บริการ	แย่มาก	ยกเลิก	เทียวบิน	ไม่แจ้ง	ล่วงหน้า	ค่ะ
	service	bad	cancel	flight	not inform	ahead	Sir
StopWord	บริการ	แย่มาก	ยกเลิก	เทียวบิน	ไม่แจ้ง	ล่วงหน้า	~~ค่ะ~~
removal	service	bad	cancel	flight	not inform	ahead	~~Sir~~
Negation word	บริการ	แย่มาก	ยกเลิก	เทียวบิน	~~ไม่~~แจ้ง	ล่วงหน้า	
split	service	bad	cancel	flight	~~not~~ inform	ahead	
Final Word	บริการ	แย่มาก	ยกเลิก	เทียวบิน	แจ้ง	ล่วงหน้า	
Sequence	service	bad	cancel	flight	inform	pre	

- Proper nouns were removed from the sentence because they do not convey any sentiment.
- Sentence was separated into words using Deep Cut tool.
- Stopwords were removed from each word in step 2, because they add no useful information.
- If a compound word contains negation syllables, it is split into two words. For example, the Thai word is "ไม่แจ้ง (Mị cæ̂ng)" contains the negation syllable "ไม่ (Mị)": it is separated into two words - "ไม่ (Mị)" and "แจ้ง (cæ̂ng)".

Finally, each word is matched with its synset; each synset with the relevant 5-tuple is extracted from the Thai Sentiment Resource. Some results are shown in Table 2.

Table 2. Examples of sentiment value tuples from Thai sentiment resources.

Words		Part of	Index	Pleasantness	Attention	Sensitivity	Aptitude	Polarity
Thai	Synsets	speech						
บริการ	Service	n	1	0.81	−0.92	0	0.79	P
	Facility	n	5	−0.57	0	0.50	−0.90	N
แย่มาก	Ill	r	1	−0.89	0	−0.99	−0.75	N
	Worse	r	1	−0.08	0	−0.09	−0.05	N
	Terribly	r	2	0	0	−0.71	−0.87	N

(*continued*)

Table 2. (*continued*)

Words		Part of speech	Index	Pleasantness	Attention	Sensitivity	Aptitude	Polarity
Thai	Synsets							
ยกเลิก	Cancel	v	5	−0.22	0	−0.11	−0.12	N
	Cancel	v	1	−0.22	0	−0.11	−0.12	N
	Break_down	v	1	−0.09	0	−0.05	0	N
	Undo	v	1	0.51	−0.50	0	0.68	P
	Abolish	v	1	0.05	−0.66	0	0.37	P
	Stay	v	9	0.05	0.13	−0.08	0.13	P
	Abandon	v	4	−0.99	0	0	−0.70	N
เที่ยวบิน	Flight	n	9	0	0.82	0	0.83	P
แจ้ง	Inform	v	1	0.54	0.87	0	0.84	P
	Inform	v	3	0.54	0.87	0	0.84	P
	Acquaint	v	3	0.65	−0.61	0	0.72	P
	Report	v	2	0	0	0.73	−0.80	N
ล่วงหน้า	Ahead	r	4	0.84	0	0	0	P

Part of speech: n = noun, v = verb, r = adverb

Index: Number assigned to various entries in the same synset.
In the following examples, each entry in a synset will be identified by an English label followed by a code for part of speech and an index, e.g. inform (v.03).

Table 3. Six methods for deciding the emotion of sentences.

Methods	Types of synset used				Reverse polarity on negation at		For each word, to decide polarity		How to calculate Sentic value	
	N.	V.	Adj.	Adv.	Word level	Sentence level	Use every synset	Use only one synset	Synset level	Sentic dimension level
1	/	/	/	/	/		/		/	
2	/	/	/	/	/		/		N/A	N/A
3		/			/			/		/
4			/	/	/			/		/
5	/	/	/	/		/	N/A	N/A		/
6		/	/	/	N/A	N/A		/	/	

4.2   Deciding Sentence Emotion

Some Thai words have one or more than one sentiment values. So a technique for inferring a single value is needed. We assess six methods to resolve this. Each method utilizes features as shown in Table 3. Details are discussed in this section.

Method 1: Average Synset Polarity Values for Each Word

- From the data in the Table 2, Thai expressions are often preceded by a negative word. If such a word is found, the polarity of the word is reversed.
- For each synset, four values in tuple found in the database are used to calculate a polarity value using the equation of SenticNet [1]:

$$(Plsn(c) + |Attn(c)| - |Snst(c)| + Aptt(c))/9 \qquad (1)$$

- The average polarity values of each word in the synset is assigned to each Thai word. An average above zero implies a positive polarity, otherwise it is assigned to negative polarity.
- Determine the majority polarity from all words of sentence. If there is a majority assign that polarity, but if there is a tie (same number of positive and negative), assign a negative polarity. The reason for this will be explained later.

For example, choosing **inform**, we have four synset entries: calculate polarity values using Eq. (1):

$$inform\ (v.01)(-0.54 + |-0.87| - |0| + (-0.84))/9 = -0.06 \qquad (2)$$

$$inform\ (v.03)(-0.54 + |-0.87| - |0| + (-0.84))/9 = -0.06 \qquad (3)$$

$$acquaint\ (v.03)(-0.65 + |0.61| - |0| + (-0.72))/9 = -0.08 \qquad (4)$$

$$report\ (v.02)(0 + |0| - |-0.73| + 0.80)/9 = 0.01 \qquad (5)$$

Calculate the average polarity value:

$$((-0.06) + (-0.06) + (-0.08) + (0.01))/4 = -0.05 \qquad (6)$$

The average is less than 0, so assign negative (N). Similarly, calculate average polarity values and assign emotions (P or N) for each synset in the remaining words: **service, bad, cancellation, flight** and **ahead**. Now find the majority polarity, in this case, there are six words and a tie (3 P, 3 N), so prefer N, leading to an overall negative (N).

Method 2: Decide Word Polarity From its Synsets Polarity Majority

- Reverse the polarity value of the word following a negation word, if any.
- Assign a polarity to each word by counting majority of polarity from its synsets.

  As before, if there is a tie, assign negative polarity.

- The sentence emotion is decided from majority of polarity of word sequence.
- In this case, we found a negation word, so select the word following it, inform. Switch the polarities, that is, positive becomes negative and vice versa. Now counting polarity majority for the other 5 words, resulting in 4 N and 2 P. Hence, method 2 predicts a negative (N) emotion here.

Method 3: Select a Representative Synset Entry of Verbs Utilizing Maximum Absolute Sentic Value

- From the data in Table 2, select only verb synsets.
- Reverse the polarity value of the word following a negation word, if any.
- Of all synsets of a word, choose the synset entry with a maximum absolute value among all four dimensions as a representative of the word.
- Decide the sentence emotion based on the majority polarity from all representative synsets. Again assign negative (N) if there is a tie.

In the example, select only the verbs, **cancel** and **inform**. There is a negation word preceding **inform,** so reverse its polarity. That is, (P, P, P, N) becomes (N, N, N, P). For each verb for each dimension, find the maximum absolute value. For **cancel**, 0.99, and **inform**, 0.87. Choose entries from each synset with the maximum value in that synset, For **cancel**, there in only one synset, **abandon**, but there are two synsets for **inform**, v.01 and v.03. Count the polarities from these three, here, 3 Ns, so the final result is negative (N).

Method 4: Select a Representative Synset Entry of Adjective and Adverb Utilizing Maximum Absolute Sentic Value

Select adjectives and adverbs. In the example, we have **bad** and **ahead**. Apply the same method as used in method 3. For **bad, ill (r.01)** has the maximum absolute value, 0.99, and corresponding polarity N. For **ahead**, the maximum is 0.84 and polarity P. There is no majority, so choose N.

Note that the decision to assign a negative polarity in case of a tie for method 1 to method 4 is based on our pilot experiment. That is, we selected 200 positive sentences and 200 negative sentences. Among these 400 sentences, there are 66, 56, 39 and 101 tied sentences (sentences with the same number of positive and negative polarity), using method 1, 2, 3 and 4 respectively. The experimental results showed that negative assignment in case of tied sentences gave overall higher accuracy of 85.25% comparing to the positive assignment of 69.5%.

Method 5: Calculate Four Maximum Sentic Dimensions at Sentence Level

- Take the absolute values of four Sentic dimensions of all synsets.
- Select the maximum absolute for each dimension.
- Use the obtained maximum values of each dimension to calculate polarity value of a sentence using Eq. (1).
- Compare the polarity value to a threshold, set to −0.027 (discussed in the next section). If higher or equal weight assign a positive emotion, otherwise, assign negative emotion.
- If the sentence contains a negation word, switch the sentence emotion. Note that we will switch only once for each sentence.

In the example, pleasantness, attention, sensitivity and aptitude have the maximum absolute values of 0.99, 0.92, 0.99 and 0.90, respectively, which corresponds to their real values of −0.99, −0.92, −0.99 and −0.90. Now use Eq. (1) to calculate an overall polarity $((-0.99) + |-0.92| - |-0.99| + (-0.90))/9 = -0.22$. Compare it with the threshold, -0.027. If it is less or equal, assign negative (N) emotion, otherwise, assign positive emotion to the sentence. In this example, assign negative (N) emotion. Finally, if there is a negation word, switch the emotion. Hence, the sentence is assigned positive (P) emotion (Table 4).

Table 4. Example of emotion assignment for method 6.

Words		Part of speech	Pleasantness	Attention	Sensitivity	Aptitude	Polarity	Polarity Value
Thai	Synsets							
	ill	r	-0.89	0	-0.99	-0.75	N	-0.29
แย่มาก	worse	r	-0.08	0	-0.09	-0.05	N	-0.02
	terribly	r	0	0	-0.71	-0.87	N	-0.18
	cancel	v	-0.22	0	-0.11	-0.12	N	-0.05
	cancel	v	-0.22	0	-0.11	-0.12	N	-0.05
	break_down	v	-0.09	0	-0.05	0	N	-0.02
ยกเลิก	undo	v	0.51	-0.50	0	0.68	P	0.19
	abolish	v	0.05	-0.66	0	0.37	P	0.12
	stay	v	0.05	0.13	-0.08	0.13	P	0.03
	abandon	v	-0.99	0	0	-0.70	N	-0.19
	inform	v	0.54	0.87	0	0.84	P	0.25
แจ้ง	inform	v	0.54	0.87	0	0.84	P	0.25
	acquaint	v	0.65	-0.61	0	0.72	P	0.22
	report	v	0	0	0.73	-0.80	N	-0.17
ล่วงหน้า	ahead	r	0.84	0	0	0	P	0.09
Average			0.09	**0.21**	-0.17	0.15		

Method 6: Select the Highest Synset Polarity Value as a Representative of Verb, Adjective and Adverb Word

- Select only verb, adjective and adverb synsets.
- Calculate the polarity value of each synset using Eq. (1).
- For each word, choose only synset(s) with the highest absolute polarity value.
- If all words have the same polarities, assign that polarity.

Otherwise,

- If words have different polarities, calculate four average values of four dimensions.
- Among the four values above, take the maximum absolute value to compare to the threshold set to 0.43 (discussed in the next section). If it is above, return a positive, otherwise return a negative.

In the example, select synsets for verbs, adjective and adverbs, we have **bad, cancel, inform** and **ahead**. Calculate polarity values from Eq. (1). Choose **ill** (r.01),

undo (v.01), **abandon** (v.04), **inform** (v.01), **inform** (v.03), and **ahead** (r.04) as representative synsets. Since their polarities are not matched, calculate four average dimension values, which are 0.09, 0.21, −0.17 and 0.15. Then decide a sentence emotion based on the maximum absolute value 0.21, which is less than the 0.43 threshold. Hence, the decision is negative (N).

The thresholds set in methods 5 and 6 were determined empirically. We selected values, in the range (−3.0, +3.0), that produced the best match with ground truths (human assigned emotions) in our 1,153 sentence database.

5 Result

The results reported from these 1,153 sentences. Table 5 reports the overall accuracy for the 401 positive and 752 negative polarities for each method. The emotions tagged by our methods were compared with sentences tagged by a human. The accuracy is reported as fraction of the correct (human assigned) result.

Table 5. Accuracy of each method.

Method	Correct tagging				Accuracy (%)
	Positive	(%)	Negative	(%)	
1	349	87	433	57	67.8
2	334	83	422	56	65.6
3	282	70	453	60	63.7
4	281	70	568	75	73.6
5	308	76	438	58	64.7
6	293	73	533	70	71.6
Majority vote	**321**	**80**	**542**	**72**	**74.9**

Table 5 shows the accuracy for each method. The accuracy calculated using equation below:

$$accuracy = \frac{(Positive + Negative) * 100}{dataset} \tag{7}$$

Method 4 showed the best accuracy. It correctly tagged 281 positive sets and 568 negative sets. The accuracy was computed from ((281 + 568) * 100)/1,153 = 73.6%. It was better than method 3 because it used adjective and adverb synsets while method 3 used only verb synsets. In general, adjective and adverb are used to express emotion more than verb. Since the accuracy from all methods was not very large, we hypothesized that a majority vote from all methods would lead to a better result. The majority vote is calculated from correct tagged result of all 6 methods. The negative emotion is assigned if there is the same count of positive and negative numbers (positive = negative). It is shown as a **Majority Vote** row in Table 5. This lead to a slightly higher success rate at 74.9%.

6 Conclusion

We evaluated six methods for tagging sentiment to Thai sentences using the previously synthesized Thai sentiment resource. Since Thai is written in the form of continuous characters without spaces and can be very ambiguous, various preprocessing steps must be applied. Those were proper noun elimination, word segmentation, stop-words removal and negation word marking, leading to a series of key words from the original sentence. Using the Thai sentiment resource, each word was associated with its (possibly multiple) synsets and a corresponding four sentic values. Then the steps of each method are applied to determine the sentence's sentiment emotion.

The test data was obtained from social media comments of an airline service. Each sentence in the dataset was manually tagged. Using a combination of the six methods, we achieved 74.9% accuracy. The best accuracy of 73.6% was achieved using only adverb and adjective synsets and the worst accuracy of 63.7% is obtained when use only verb synsets.

References

1. Cambria, E., Speer, R., Havasi, C., Hussain, A.: SenticNet: a publicly available semantic resource for opinion mining. In: AAAI Fall Symposium: Commonsense Knowledge, vol. 10, no. 0, pp. 14–18 (2010)
2. Lertsuksakda, R., Netisopakul, P., Pasupa, K.: Thai sentiment terms construction using the Hourglass of Emotions. In: 2014 6th International Conference on Knowledge and Smart Technology (KST), pp. 46–50. IEEE (2014)
3. Netisopakul, P., Thong-iad, K.: Thai sentiment resource using Thai WordNet. In: Conference on Complex, Intelligent, and Software Intensive Systems, pp. 329–340, July 2018
4. Thoongsup, S., Robkop, K., Mokarat, C., Sinthurahat, T., Charoenporn, T., Sornlertlamvanich, V., Isahara, H.: Thai wordnet construction. In: Proceedings of the 7th Workshop on Asian Language Resources. Association for Computational Linguistics, pp. 139–144 (2009)
5. Cambria, E., Poria, S., Bajpai, R., Schuller, B.: SenticNet 4: a semantic resource for sentiment analysis based on conceptual primitives. In: Proceedings of COLING 2016, The 26th International Conference on Computational Linguistics: Technical Papers, pp. 2666–2677 (2016)
6. Lertsuksakda, R., Pasupa, K., Netisopakul, P.: Sentiment analysis of Thai children stories on support vector machine. In: 2015 Proceeding of the Twentieth International Symposium on Artificial Life and Robotics (AROB), Beppu, Japan, pp. 138–142 (2015)
7. Havasi, C., Alonso, J., Speer, R.: ConceptNet 3: a flexible, multilingual semantic network for common sense knowledge. In: Recent Advances in Natural Language Processing, pp. 27–29. John Benjamins, Philadelphia (2007)
8. NECTEC: Lexitron Dictionary. http://lexitron.nectec.or.th
9. Belisan: Volubilis Dictionary (2005–2014). http://belisan-volubilis.blogspot.com
10. Kittinaradorn, R., Chaovavanich, K., Achakulvisut, T., Kaewkasi, Ch.: Deepcut (2008). https://github.com/rkcosmos/deepcut

Collecting Child Psychiatry Documents of Clinical Trials from PubMed by the SVM Text Classification Method with the MATF Weighting Scheme

Jantima Polpinij[1]([⊠]), Tontrakant Kachai[1], Kanyarat Nasomboon[1],
and Poramin Bheganan[2]([⊠])

[1] Intellect Laboratory, Department of Computer Science, Faculty of Informatics,
Mahasarakham University, Mahasarakham Province, Thailand
{jantima.p,tontrakant.kac,kanyarat.nas}@msu.ac.th
[2] Computer Science Program, Mahidol University International College,
Mahidol University, Nakhonpathom Province, Thailand
poramin.bhe@mahidol.ac.th

Abstract. Child psychiatry is a branch of psychiatry focused on the diagnosis, treatment, and prevention of mental health issues in children and their families. In many countries, the study of disorders such as ADHD (Attention-Deficit/Hyperactivity Disorder) by child and adolescent psychiatry is still in its infancy, with the result that children's mental health issues can be the source of embarrassment for the family and of shame for many children. Misunderstanding, denying, and ignoring children's mental health issues by parents are the main problem encountered in diagnosis and treatment of mental health issues in children. To help parents and extended families understand this problem better, and thus help them to better care for children with mental health issues, starting with seeking help from a psychiatrist without embarrassment, an easily accessible and reliable source of information is urgently needed. To develop such a single source of information, relevant documents need to be gathered together. This study presents a method of gathering reports of clinical trials from PubMed which describe diagnosis and treatment of child mental health issues. The main mechanism of the proposed method is a Support Vector Machine with a Multi Aspect TF (MATF) weighting scheme. After testing by recall, precision, and F1, it can return satisfactory results of 0.82, 0.79, and 0.80 respectively.

Keywords: Child psychiatry · Clinical trials · ADHD ·
Text classification · Support Vector Machines · Multi Aspect TF ·
PubMed

1 Background

Child psychiatry, a branch of psychiatry, emphasizes the diagnosis, treatment, and prevention of mental disorders in children and their families [1–3]. According

P. Boonyopakorn et al. (Eds.): IC2IT 2019, AISC 936, pp. 99–108, 2020.
https://doi.org/10.1007/978-3-030-19861-9_10

to child psychiatry, various types of mental health disorders can affect children such as Attention-Deficit/Hyperactivity Disorder (ADHD), Learning Disorder (LD), depression, bipolar and disruptive behavior disorders, eating disorders, and so on. ADHD is a behavioral disorder such that children with ADHD are excessively active, unable to concentrate, or both, and so ADHD impacts children's learning, academic performance, and their daily life. Signs of the symptoms usually appear before the age of 7 years. Studies have shown that boys are more likely to be diagnosed with ADHD than girls. In addition, half of all children with ADHD also have a LD. As a consequence ADHD has become the most common mental health issue found in children [1–7].

To the best of our knowledge, it has been revealed that more boys than girls are diagnosed with ADHD. In addition, half of all children with ADHD also have a learning disorder (LD). Then, ADHD become the most common mental health problems found in children [4–7].

However, different countries have different environments, so the background of ADHD including its diagnosis and treatment should be different as well. What we do know is that without sufficient knowledge of ADHD, parents and families are likely to misunderstand their children's mental health issues, and so they may deny or ignore them. Even today ADHD is still a problem in many countries since children's mental health issues are considered a source of embarrassment for the family and of shame for many children [4–7]. To address this issue, an easily accessible and reliable source of knowledge for parents and families is crucial. An excellent source of ADHD knowledge is PubMed, as are many mental health related web sites.

However, mental health information is not easy to find and capture in PubMed, while mental health related web sites might contain only general information about ADHD. Therefore, a website containing detailed information about ADHD is required. A first step in its development must be to collect relevant documents with ADHD knowledge. This is a challenge for this study where we aim to present a method of automatically gathering documents from the field of child psychiatry from *PubMed*, where those documents reference ADHD knowledge in clinical trials and are presented in textual form. The proposed method makes use of natural language processing (NLP) and text mining (TM).

The rest of the paper is organized as follows: Sect. 2 presents the dataset; Sect. 3 the research method; Sect. 4 the experimental results and discussion; and Sect. 5 the conclusion.

2   Related Works

Many studies have proposed methods for extracting knowledge related to issues in the medical domain, and some are briefly reviewed below.

Zhu et al. [8] applied the apriori association rule mining algorithm to generate the co-occurrences of medical concepts, which are then filtered through a set of predefined semantic templates to instantiate useful relations. From such semantic relations, decision elements and possible relationships among them may be derived for the construction of a clinical decision model. To evaluate the proposed method, they conducted a case study in colorectal cancer management, and preliminary results showed that useful causal relations and decision alternatives could be extracted.

In the last decade, it has been found that NLP and TM are regularly applied to extract important knowledge from medical documents stored in MEDLINE. Some examples can be found in Nuzzo et al. [9] and Miñarro-Giménez et al. [10].

Nuzzo et al. [9] developed a knowledge extraction tool to help researchers in discovering useful information which could support their reasoning process. The tool is composed of a search engine based on TM and NLP techniques, and an analysis module which processes the search results to build annotation similarity networks. They tested our approach on the available knowledge about the genetic mechanism of cardiac diseases, where the target is to find both known and possible hypothetical relations between specific candidate genes of interest. They also showed that the system is able to effectively retrieve medical concepts and genes and it plays a role in assisting researchers in the formulation and evaluation of novel literature-based hypotheses

Miñarro-Giménez et al. [10] explored the combination of a clustering method for co-occurring concepts based on their related MeSH subheadings in MEDLINE with the use of SemRep, an NLP engine, which extracts predications from free text documents. As a result, they generated sets of clusters of co-occurring concepts and identified the most significant predicates for each cluster. The association of such predicates with the co-occurrences of the resulting clusters produced the list of predications, which were checked for relevance.

Some studies have sought to find or collect knowledge related to child psychiatry in clinical trials from medical textual data. Such knowledge presents the body of knowledge related to diagnosis and treatment processes. Some of these studies are briefly reviewed below.

Karystianis et al. [11] studied domestic violence (DV), and explored whether text mining can automatically identify mental health disorders from unstructured text related to DV. After testing by test set, the precision returned from an evaluation set of 100 DV events was 0.975 and 0.871 for mental health disorders related to person of interest (POIs) and victims, respectively. When they applied their proposed method to a larger corpus, it returned satisfactory results. Therefore, they demonstrated that text mining could automatically extract required information from police-recorded DV events, which will help to support further studies of mental health disorders and DV.

Leroy et al. [12] mentioned that electronic health records (EHR) benefit in information utilization, e.g. in the surveillance conducted by the Centers for Disease Control and Prevention (CDC) to track cases of autism spectrum disorder (ASD). Their objective was to automatically leverage the description of behaviors in EHRs made by clinicians regarding the diagnostic criteria in the Diagnostic and Statistical Manual of Mental Disorders (DSM). After first studying the classification of ASD in the entire EHRs, they focused on the extraction of individual expressions of the different ASD criteria in the text. They also intended to facilitate extensive observations of ASD, strengthen the analysis of changes over time, and integrate these with other relevant data.

3   The Dataset

Our dataset is from PubMed, and the documents in this dataset are related to ADHD. We use only research abstracts for our study. The documents in the dataset were downloaded between 1 and 31 August 2016. The dataset consists of 600 documents divided into two groups: general and clinical trials. The documents are formatted as text files, and are different in length. The training set contains 400 documents (200 general abstracts and 200 clinical trial abstracts), and the test set contains 200 reviews (100 general abstracts and 100 clinical trial abstracts). It is noted that this work concentrates on reviews written in English. An example is shown as Fig. 1.

OBJECTIVE: To study the clinical characteristics of children who were diagnosed as ADHD.

MATERIAL AND METHOD: A retrospective chart review was conducted on 202 children who came to a child mental health clinic and were diagnosed as ADHD.

RESULTS: Most cases were in the 6-12 years age group and came from small families with 1-2 children. Males outnumbered females (M:F = 3.4:1). One-fifth of the sample received previous psychiatric evaluation from other health professionals but parents needed 'second opinion'. The most frequent chief complaints were academic/learning problems. Almost one-fourth of the samples came for problems

not directly related to ADHD. In this group the most frequent complaints were aggressive and oppositional behavior Comorbidity was found in 53.5%. More than half of the cases who took intelligence tests had an IQ below 90. Behavioral management was the only treatment modality in 38% of the sample. In 62% stimulants were instituted either at the beginning of treatment or as an "add-on" after behavioral management proved to be insufficient. Among cases that received

stimulants, 28% needed the combination of other psychotropic medications, mostly antidepressant and anxiolytic drugs.

CONCLUSION: A study of the clinical characteristics of ADHD in Thai children revealed male preponderance and high rates of non-ADHD presentations and comorbid conditions. Awareness of varied presentations of ADHD and proper treatment of comorbid conditions is imperative in the comprehensive care of ADHD children.

Fig. 1. A PubMed abstract related to ADHD

4   The Research Method

This section presents the research method, and explains the proposed method for modelling text classifiers. In general, the goal is to automatically classify the text documents into one or more defined classes. Suppose D is set of documents, denoted as $D = \{d_1, d_2, \cdots, d_n\}$. Meanwhile, a fixed set of classes is denoted as $C = \{c_1, c_2, \cdots, c_j\}$. Therefore, a predicted class can be $c \in C$. In this study, C is the set of two classes {clinical trial, general}.

4.1 Word Segmentation, Feature Representation, and Term-Word Weighting

Firstly, the training set separates text into words using word delimiters (i.e. white space), and then stop-words will be removed. After those processes are carried out, each document and its features are represented in the format of a *vector space model*, denoted as $D = \{d_1, d_2, d_3, \cdots, d_n\}$. The alternative name of this structure is *bag of words (BOW)*. In BOW, each term-word is weighted by MATF (Multi Aspect TF), which was proposed by Paik [13]. This weighting scheme is a modification of *tf-idf*, where term frequency *(tf)* is effective for short documents, while the other performs better on long documents. Therefore, the aim of this algorithm is to balance the various numbers of features in documents. We calculate the MATF with the following equation:

$$MATF(t, D) = TFF(t, D) \times TDF(t, C) \tag{1}$$

where t is a concept found in a document D, while C is a document corpus. Transforming *TF* Factors *(TFF)* is the local weighting and the *Term Discrimination Factor (TDF)* is the global weighting. The formula for TFF is:

$$TFF(t, D) = w \times BRITF(t, D) + (1 - w) \times BLRTF(t, D) \tag{2}$$

The value of w is then calculated with the following formula:

$$w = \frac{2}{1 + log_2(1 + |Q|)} \tag{3}$$

BRITF is to calculate short length documents, using the following formula:

$$BRITF(t, D) = \frac{RITF(t, D)}{1 + RITF(t, D)} \tag{4}$$

The formula for calculating *RITF (Relative intra-document)* is:

$$RITF(t, D) = \frac{log_2(1 + TF(t, D))}{log_2(1 + Avg.TF(t, D))} \tag{5}$$

where *TF(t, D)* is the frequency of the concept t found in document D, and *Avg. TF(t, D)* is the average of the frequency of all concepts found in document D. For *BLRTF*, it is used to calculate the long length text of the customer review. The formula is,

$$BLRTF(t, D) = \frac{LRTF(t, D)}{1 + LRTF(t, D)} \tag{6}$$

To calculate *LRTF (Length regularized TF)* value, we use the following formula:

$$LRTF(t, D) = TF(t, D) \times log_2 \left(1 + \frac{ADL(C)}{len(D)}\right) \tag{7}$$

where $ADL(C)$ is the average of customer review length in the corpus, and $len(D)$ is the length of a customer review. To calculate TDF, we use the following formula:

$$TDF(t, D) = IDF(t, C) \times \frac{AEF(t, C)}{1 + AEF(t, C)} \tag{8}$$

$$IDF(t, C) = log\left(\frac{CS(C) + 1}{DF(t, D)}\right) \tag{9}$$

where $CS(C)$ is the total number of customer reviews in the corpus, and $DF(t, C)$ is the total number of customer reviews containing concept t. Meanwhile, $AEF(t, C)$ is,

$$AEF(t, C) = \frac{CTF(t, C)}{DF(t, C)} \tag{10}$$

where $CTF(t, C)$ is the total time of concept t is found in the corpus, and $DF(t, C)$ is the number of customer reviews containing concept t.

4.2   Text Classifier Modelling

The study chooses support vector machines (SVM) to model text classifiers because the SVM can work effectively not only with document vectors with sparse features [14, 15] but also to perform well with imbalanced data [16]. In this stage, we develop text classifiers by the SVM algorithm, where the BOW with the MATF weighting scheme is obtained from the previous stage. The SVM is chosen because the basic concept behind the training procedure is to find a hyperplane, represented by vector w, that not only separates the document vectors in one class from those in the other, but for which the separation, or margin, is as large as possible. This search corresponds to a constrained optimization problem. Let $c_j \in \{1, -1\}$ be the correct class of document d_j. The study concentrates only two classes: positive and negative classes. The solution can be:

$$Minimize : v(w, \varepsilon, \rho) = \frac{\|w\|^{-2}}{2} + \frac{1}{vl}\sum_{i=1}^{l}\varepsilon_i - \theta \tag{11}$$

$$Subject\ to : w \cdot \lambda \geq \rho - \varepsilon_i, \varepsilon_i \geq 0 \tag{12}$$

where $\nu \in (0, 1)$ is a parameter, which let one controls the number of support vectors and errors, ε is a measure of the misclassification errors, and ρ is the margin. If we solve the problem, we can obtain w and ρ. Given a new data point x to classify, a label is assigned according to the decision function that can be expressed as follows:

$$f(x) = sign(w \cdot \lambda(x_i) - \rho) \tag{13}$$

We can set the derivatives with respect to the primal variables to equal zero, and then we get:

$$W = \Sigma\alpha_i \times \lambda(x_i) \tag{14}$$

where α_i are the Lagrange multipliers and we apply the Kuhn Tucker condition.

There is only one subset of points x_i that lies closest to the hyperplane and has nonzero values α_i. These points are called support vectors. Instead of solving the primal optimization problem directly, the dual optimization problem is given by:

$$Minimize : W(\alpha) = \frac{1}{2} \sum_{i,j} \alpha_i \alpha_j K(x_i, x_j) \tag{15}$$

$$Subject\ to : 0 \le \alpha_i \le \frac{1}{vl}, \sum_i \alpha_i = 1 \tag{16}$$

where $K(x_i, x_j)$ is the kernel function performing the non-linear mapping into feature space based on dot products between mapped pairs of input points. Finally, we will get text classifiers used to classify documents into two defined classes.

5   The Experimental Results

The common performance measures for information system evaluation are applied. They are *recall (R)* [17], *precision (P)* [17], and *F-measure (F1)* [17]. These are illustrated by the confusion matrix shown in Table 1. By using the confusion matrix, *recall*, *precision*, and *F1* can be calculated as follows.

$$R = \frac{tp}{tp + fn} \tag{17}$$

$$P = \frac{tp}{tp + fp} \tag{18}$$

$$F1 = 2 \times \frac{R \times P}{R + P} \tag{19}$$

Table 1. The experimental results

		Predicted	
		Clinical trial	General
Actual	Clinical trial	true positive (tp)	false negative (fn)
	General	false positive (fp)	true negative (tn)

However, we also compare our proposed method to the traditional text classification, where the traditional text classifier modelling often uses *tf* and *tf-idf* as weighting schemes. The results of the comparison are shown in Table 2. By using the MATF weighting scheme, it can be seen that our proposed method is better than the traditional text classification method.

Consider the results in Table 2. It can be seen that the text classifier model with the MATF weighting scheme returns better results than the text classifier

Table 2. The experimental results

Methods	Recall	Precision	F-measure
Text classifier $(tf + SVM)$	0.78	0.75	0.76
Text classifier $(tf - idf + SVM)$	0.75	0.73	0.74
Text classifier $(MATF + SVM)$	0.82	0.79	0.80

model with *tf (term frequency)* weighting scheme and the text classifier model with *if-idf (term frequency – inverse document frequency)* weighting scheme. This is because the MATF can handle two different within document term frequency normalizations to capture two different aspects of term saliency. One component of the term frequency is effective for short documents, while the other performs better on long documents. The final weight is then measured by taking a weighted combination of these components, which is determined on the basis of the length of the corresponding documents. Simply speaking, the MATF can handle the problem of imbalanced data better than *tf* and *tf-idf*. Meanwhile, *tf* is better than *tf-idf*, because terms such as "ADHD", "treatment", or "diagnosis" actually appear in many documents but, as we discovered, constitute interesting features that guide the classification techniques to distinguish clinical trials documents from general documents.

In addition, it is possible that documents used in this study can be of very different sizes. Therefore, the classes may not contain the term-words equally. Simply speaking, the number of features belonging to one class (called minority class) is significantly lower than those belonging to the other classes (called majority class). Without doubt, the document classification decision tends to be overwhelmed by the majority class and ignores the minority class if using traditional classifiers. Therefore, the SVM algorithm is chosen for this study, since it has been confirmed by Akbani et al. [16] that this algorithm can perform well for a dataset with an imbalanced data problem.

However, our proposed method returns satisfactory outputs when used to gather documents related to ADHD knowledge in clinical trials from PubMed. After testing the proposed method via recall, precision, and F1, the results are satisfactory, where the scores of recall, precision, and F1 are 0.82, 0.79, and 0.80 respectively.

6   Conclusion

Child psychiatry confronts one of the problems that requires attention today. Frequently parents and families have inadequate knowledge of their children's mental health. Moreover, they may feel embarrassed about their children's mental health issues. As a consequence many families may ignore and deny the problems. Yet provided the children are treated and looked after appropriately, they can function normally in society. PubMed and mental health related web sites are important sources of ADHD knowledge, yet such information in PubMed is never easy to find and extract, whereas mental health related web sites might contain only general information about ADHD. Therefore, if a website containing essential information about ADHD were developed, with the information made easily accessible, it would be of great benefit to the parents and families of children with ADHD. However, a collection of relevant documents with ADHD knowledge should be made first, and it is to further this goal that the current study presents a method to retrieve and collect data from PubMed regarding ADHD in clinical trials. Those documents could then be used in the construction of a knowledge base.

References

1. Polanczyk, G., de Lima, M.S., Horta, B.L., Biederman, J., Rohde, L.A.: The world-wide prevalence of ADHD: a systematic review and metaregression analysis. Soc. Hist. Med. **164**(6), 942–948 (2007)
2. Smith, M.: Hyperactive around the world? The history of ADHD in global perspective. Soc. Hist. Med. **30**(4), 767–787 (2017)
3. Davidovitch, M., Koren, G., Fund, N., Shrem, M., Porath, A.: Challenges in defining the rates of ADHD diagnosis and treatment: trends over the last decade. BMC Pediatr. **17**, 218 (2017)
4. Trangkasombat, U.: Clinical characteristics of ADHD in Thai children. J. Med. Assoc. Thai. **91**(12), 1894–1898 (2008)
5. Zheng, Y., Zheng, X.: Current state and recent developments of child psychiatry in China. Child Adolesc. Psychiatry Ment Health **9**, 10 (2015)
6. Hirota, T., Guerrero, A., Sartorius, N., Fung, D., Leventhal, B., Ong, S.H., Kaneko, H., Apinuntavech, S., Bennett, A., Bhoomikumar, J., Cheon, K.A., Davaasuren, O., Gau, S., Hall, B., Koren, E., van Nguyen, T., Oo, T., Tan, S., Tateno, M., Thikeo, M., Wiguna, T., Wong, M., Zheng, Y., Skokauskas, N.: Child and adolescent psychiatry in the Far East: a 5-year follow up on the Consortium on Academic Child and Adolescent Psychiatry in the Far East (CACAP-FE) study. Psychiatry Clin. Neurosci. **73**(2), 84–89 (2018)
7. Pejovic-Milovancevic, M., Miletic, V., Anagnostopoulos, D., Raleva, M., Stancheva, V., Burgic-Radmanovic, M., Barac-Otasevic, Z., Ispanovic, V.: Management in child and adolescent psychiatry: how does it look in the Balkans? Psychiatriki **25**(1), 48–54 (2014)
8. Zhu, A.L., Li, J., Leong, T-Y.: Automated knowledge extraction for decision model construction: a data mining approach. In: The AMIA Annual Symposium Proceedings, pp. 758–762 (2003)
9. Nuzzo, A., Mulas, F., Gabetta, M., Arbustini, E., Zupan, B., Larizza, C., Bellazzi, R.: Text Mining approaches for automated literature knowledge extraction and representation. Stud. Health Technol. Inf. **160**(Pt 2), 954–958 (2010)
10. Miñarro-Giménez, J.A., Kreuzthaler, M., Schulz, S.: Methods for knowledge extraction from a clinical database on liver diseases. In: The AMIA Annual Symposium Proceedings, pp. 915–924 (2015)
11. Karystianis, G., Adily, A., Schofield, P., Knight, L., Galdon, C., Greenberg, D., Jorm, L., Nenadic, G., Butler, T.: Automatic extraction of mental health disorders from domestic violence police narratives: text mining study. J. Med. Internet Res. **20**(9), e11548 (2018)

12. Leroy, G., Gu, Y., Galindo, M.K., Arora, A., Kurzius-Spencer, M.: Automated extraction of diagnostic criteria from electronic health records for autism spectrum disorders: development, evaluation, and application. J. Med. Internet Res. **20**(11), e10497 (2018)
13. Paik, J.H.: A novel TF-IDF weighting scheme for effective ranking. Crit. Rev. Anal. Chem. **36**(1), 41–59 (2013)
14. Joachims, T.: Text categorization with support vector machines: learning with many relevant features. In: European Conference on Machine Learning (ECML), pp 137–142 (1998)
15. Joachims, T.: Transductive inference for text classification using support vector machines. In: Proceedings of the International Conference on Machine Learning (ICML) (1999)
16. Akbani, R., Kwek, S., Japkowicz, N.: Applying support vector machines to imbalanced datasets. In: European Conference on Machine Learning, pp. 39–50 (2004)
17. Baeza-Yates, R., Ribeiro-Neto, B.: Modern Information Retrieval. ACM Press, New York (1999)

Image Processing

Instance-Based Learning for Blood Vessel Segmentation in Retinal Images

Worapan Kusakunniran[✉], Sarattha Kanchanapreechakorn,
and Kittikhun Thongkanchorn

Faculty of Information and Communication Technology,
Mahidol University, Nakhon Pathom 73170, Thailand
worapan.kun@mahidol.edu, sarattha.kar@student.mahidol.ac.th,
kittikhun.tho@mahidol.ac.th
https://sites.google.com/a/mahidol.edu/worapan-kusakunniran/

Abstract. Diabetic retinopathy is a fatal disease that affects the majority of those who have diabetes for a period of time. It could lead to a blindness in the end. Therefore, it is important to detect the diabetic retinopathy in the early stage, in order to prevent the blindness. One of the key indicators of the disease is the abnormality of blood vessels in the retina. This research paper is thus to propose the technique to automatically segment blood vessels in retinal images, which could be used further for the disease analysis. It begins with using the color transfer approach to normalize the color statistics of all input images based on the reference image in the lab color space. Then, the magenta channel is extracted and used for the best distinct of blood vessel structure from the background. The morphological operators and binarization process are applied here for segmenting the blood vessels with the noise reduction using CLAHE or contrast limited adaptive histogram equalization. The proposed method is validated using the published dataset, namely STARE. The proposed method achieves the promising sensitivity and specificity.

Keywords: Diabetic retinopathy · Vessel segmentation ·
Retinal image

1 Introduction

This paper is to propose a technique of vessel segmentation in retinal images. The segmentation output from the proposed approach could be used for a further analysis of the eye disease detection, particular for stages 3 and 4 of diabetic retinopathy that focus on abnormality of blood vessels in the retina. The existing methods of vessel segmentation are roughly categorized into two types which are supervised-based learning and instance-based learning approaches.

In the supervised-based learning approaches, the pixel values of vessels of the training retinal images are fed in the learning process, producing the model(s) of

© Springer Nature Switzerland AG 2020
P. Boonyopakorn et al. (Eds.): IC2IT 2019, AISC 936, pp. 111–118, 2020.
https://doi.org/10.1007/978-3-030-19861-9_11

vessel pixel-based classification. For example, in [1], the support vector machine (SVM) was employed as the main learning tool. It applied on features extracted from the basic and orthogonal line operators in twelve different angles. The green channel was also used as a part of the features. Similarly, the method proposed in [2] also used the SVM and Bayesian as the main vessel classifiers. It applied on seven features extracted based on Gabor wavelet, line operator, and inverted green channel.

The following two examples of exiting methods also relied on the SVM as the classification toot, in order to find the optimal hyperplane for segmenting vessels from backgrounds. The method proposed in [3] applied morphological operators before the feature extraction process. The opening operation was applied to remove smaller bright details of non-blood vessels. They could be noises or pathological areas. The closing operation was used to enhance dark details. They mainly contained vessels. Then, the difference between erosion and dilation was the enhancement of the edge information. Later, the local area shape-based features were extracted and combined with the multi-scale local statistical features from the green channel. While, in [4], on top of the morphological operations, the adaptive histogram equalization was applied in the pre-processing step. The features were extracted based on the local binary pattern and the gray level co-occurrence matrix, using energy, contrast, correlation, and homogeneity.

In addition, the method proposed in [5] used the cellular neural networks as the classification tool. The features were extracted based on the line detection on the inverted green channel. It was shown to be effective in the challenges noises and low contrast between vessels and backgrounds. In [6], the neural network was also used as the pixel-based classification tool. The 7D-feature vector was computed in the gray-level-based and moment invariants-based feature spaces. The pixel-gaps filtering and false detection removal were applied in the post-processing step.

The key challenge of the supervised-based learning approaches is that it could easily stuck in the overfitting when there is not sufficient number of training samples. This occurs often in this area of medical image analysis, when the large number of medical images with related groundtruth is difficult to be obtained.

To overcome this limitation, the instance-based learning approaches were introduced for the vessel segmentation. For example, in [7], the vessels were detected by approximately forming up straight-line elements which were grouped from sets of primitives. These primitives referred to ridges extracted from retinal images. In addition, the self-organizing map (SOM) with the Otsu's method were applied for the vessel segmentation in the enhanced retinal images [8].

Differently in [9], it used Heidelberg retina tomography (HRT) images which contained typical double-edge shapes of retinal vessels with different qualities. The vascular structures reconstruction was performed on the green channel. The Gaussian function was used to detect the center line of vessels with the certain widths. Then, the unbiased detector of curvilinear structures was applied to detect vessels.

In [10], the iterative graph cut was used to segment the vessel pixels from the backgrounds. The graph was constructed using the directions and magnitudes of the vessels' gradients. These pieces of information was extracted in the distance map image generated using the distance transform algorithm. The histogram equalization was also used to enhance the contrast of vessels in the pre-processing step.

The key challenge of the instance-based learning approaches is the intra-variation of color tones and intensities of vessels, among retinal images. This makes the method sensitive to different input images, especially when they are captured from different sources.

In this paper, such challenge will be reduced using the color transfer technique. It could normalize the color statistics of input retinal images, before the further analysis of the vessel segmentation.

The rest of this paper is organized as follows. The proposed method is explained in the Sect. 2. The experimental results are shown in the Sect. 3. Then, the conclusion is drawn in the Sect. 4.

2   Proposed Method

The framework of the proposed method of vessel segmentation is illustrated in Fig. 1. The input retinal images are originally loaded in the RGB color model. They contains intra-variations particularly on their color statistics. This makes it difficult for the further analysis of the parameters tuning in the proposed instance-based learning approach.

Therefore, the input images are first normalized with the color statistic of the reference image, using the color transfer technique [11]. In this paper, one good quality retinal image is selected as used as the reference image (I_R). Both input retinal image (I_1) and I_R are transformed from the RGB color model into the $l\alpha\beta$ color model [12]. It operates in this color space because of the decorrelation of the three channels which allows the simplification of the method to treat each channel independently.

Then, the means and standard deviations of both images are calculated for each color channel separately. These color statistics of I_1 are normalized to match with the color statistics of I_R, by applying the following calculations to each pixel in each color channel.

$$\hat{l}_p = (l_p - \mu_l^1)\frac{\sigma_l^R}{\sigma_l^1} + \mu_l^R$$

$$\hat{\alpha}_p = (\alpha_p - \mu_\alpha^1)\frac{\sigma_\alpha^R}{\sigma_\alpha^1} + \mu_\alpha^R \qquad (1)$$

$$\hat{\beta}_p = (\beta_p - \mu_\beta^1)\frac{\sigma_\beta^R}{\sigma_\beta^1} + \mu_\beta^R$$

where l_p, α_p and β_p are values of pixels p^{th} in l, α and β channels respectively, $\hat{l}_p$, $\hat{\alpha}_p$ and $\hat{\beta}_p$ are the color-transferred values of pixels p^{th} in l, α and β channels

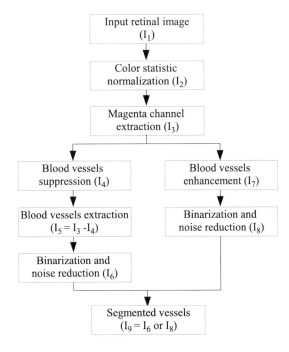

Fig. 1. The overview of the proposed framework.

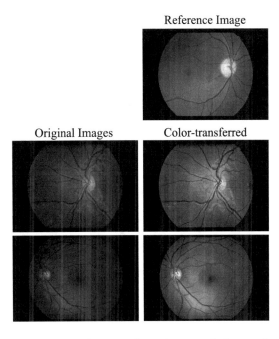

Fig. 2. Sample outputs of the color transfer, using sample images in the e_ophtha EX dataset.

Table 1. The experimental results and comparisons.

Methods	Sensitivity	Specificity
[13]	75.2	96.8
[14]	75.8	95.0
[15]	61.3	97.6
[16]	73.1	96.8
[3]	70.4	98.7
[17]	67.5	95.7
[18]	73.1	96.8
[19]	57.0	91.5
[6]	69.4	98.2
[20]	75.1	95.7
[21]	70.0	97.3
[22]	71.7	97.5
[7]	69.7	98.1
[23]	72.6	97.6
[24]	71.8	97.5
The proposed method	78.6	91.2

respectively, μ_l^1, μ_α^1, μ_β^1 are mean values of I_1 in l, α and β channels respectively, μ_l^R, μ_α^R, μ_β^R are mean values of I_R in l, α and β channels respectively, σ_l^1, σ_α^1, σ_β^1 are standard deviation values of I_1 in l, α and β channels respectively, and σ_l^R, σ_α^R, σ_β^R are mean values of I_R in l, α and β channels respectively.

Once the image I_1 is color-transferred using the Eq. (1), it is transformed back into the RGB color space (I_2). The sample color-transferred outputs are shown in Fig. 2.

The next step is to extract only the magenta channel (I_3) from I_2 for the further analysis [25]. This I_3 is used for the vessel segmentation in two manners. In the first manner, the vessels are suppressed by applying the opening operation [26] on I_3 to generate I_4. Then, I_4 is subtracted from I_3 to compute I_5 which remains with the majority of vessel pixels. The image thresholding [27] and erosion operation [28] are applied on I_5 to compute the first version of segmented vessels (I_6), by performing the binarization and noise removal respectively.

In the second manner, instead of suppressing the vessels, vessels in the magenta channel (I_3) are enhanced using the contrast limited adaptive histogram equalization (CLAHE) [29] to generate I_7. It could enhance the local contrasts between vessels and other backgrounds. Also, the image thresholding and erosion operation are applied on I_7 to compute the second version of segmented vessels (I_8). The final segmented vessels image is computed by combining I_6 and I_8 using OR operation.

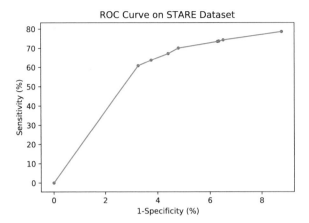

Fig. 3. ROC curve on the STARE dataset.

3   Experiments

In this paper, the proposed method is evaluated using the published dataset, namely the structured analysis of the retina (STARE) dataset [30]. It contains 20 digitized images captured by the TopCon TRV-50 fundus camera at 35° FOV, where ten of them have pathology. The images are in the ppm format, with the size of 700×605 pixels of eight bits per color channel.

The sensitivity and specificity values reported by the proposed method are compared to the other existing methods in the literature based on the same STARE dataset, as shown in the Table 1. The sensitivity is calculated by dividing the true positive with the summation of the true positive and the false negative. It represents how well the proposed method can detect the blood vessels. While, the specificity is calculated by dividing the true negative with the summation of the true negative and the false positive. It represents how well the proposed method can detect the non-blood vessels. The ROC curve trading off between the sensitivity and 1 - specificity is also shown in Fig. 3.

It can be seen that the proposed method achieves the highest value of sensitivity, when compared with the other existing methods in the literature. However, our specificity value is lower than the others. This is because the proposed method attempts to boost up the sensitivity or true positive rate, in order to detect blood vessels in the retinal image as much as possible. Therefore, they could be used in the disease analysis of the diabetic retinopathy.

4   Conclusion

This paper proposes the instance-based learning approach of the vessel segmentation in retinal images. The final segmented blood vessels are from a combination of two segmentation results. One is generated from subtracting the vessel-suppression from its original magenta channel. While, the another is generated from enhancing the contrast between vessels and other backgrounds. In addition, before performing the mentioned segmentation process, the color transfer approach is applied in the pre-processing step, to normalize the color statistics of input retinal images. The proposed method is validated using the published STARE dataset. It achieves the outstanding sensitivity value, when compared with the others in the literature.

Acknowledgments. This work was supported by the Office of Higher Education Commission (OHEC) Thailand and the Thailand Research Fund (TRF) [grant number MRG6080267].

References

1. Ricci, E., Perfetti, R.: Retinal blood vessel segmentation using line operators and support vector classification. IEEE Trans. Med. Imaging **26**(10), 1357–1365 (2007)
2. Kharghanian, R., Ahmadyfard, A.: Retinal blood vessel segmentation using gabor wavelet and line operator. Int. J. Mach. Learn. Comput. **2**(5), 593 (2012)
3. Han, Z., Yin, Y., Meng, X., Yang, G., Yan, X.: Blood vessel segmentation in pathological retinal image. In: 2014 IEEE International Conference on Data Mining Workshop (ICDMW), pp. 960–967. IEEE (2014)
4. Raja, D.S.S., Vasuki, S., Kumar, D.R.: Performance analysis of retinal image blood vessel segmentation. Adv. Comput. **5**(2/3), 17 (2014)
5. Perfetti, R., Ricci, E., Casali, D., Costantini, G.: Cellular neural networks with virtual template expansion for retinal vessel segmentation. IEEE Trans. Circ. Syst. II: Expr. Briefs **54**(2), 141–145 (2007)
6. Marín, D., Aquino, A., Gegúndez-Arias, M.E., Bravo, J.M.: A new supervised method for blood vessel segmentation in retinal images by using gray-level and moment invariants-based features. IEEE Trans. Med. Imaging **30**(1), 146 (2011)
7. Staal, J., Abràmoff, M.D., Niemeijer, M., Viergever, M.A., Van Ginneken, B.: Ridge-based vessel segmentation in color images of the retina. IEEE Trans. Med. Imaging **23**(4), 501–509 (2004)
8. Zhang, J., Cui, Y., Jiang, W., Wang, L.: Blood vessel segmentation of retinal images based on neural network. In: International Conference on Image and Graphics, pp. 11–17. Springer, Heidelberg (2015)
9. Paulus, D., Chastel, S., Feldmann, T.: Vessel segmentation in retinal images. In: Medical Imaging 2005: Physiology, Function, and Structure from Medical Images, vol. 5746, pp. 696–706. International Society for Optics and Photonics (2005)
10. Salazar-Gonzalez, A.G., Kaba, D., Li, Y., Liu, X.: Segmentation of the blood vessels and optic disk in retinal images. IEEE J. Biomed. Health Inf. **18**(6), 1874–1886 (2014)
11. Reinhard, E., Adhikhmin, M., Gooch, B., Shirley, P.: Color transfer between images. IEEE Comput. Graph. Appl. **21**(5), 34–41 (2001)
12. Bradski, G., Kaehler, A.: Learning OpenCV: Computer vision with the OpenCV library. O'Reilly Media, Inc. (2008)
13. Al-Diri, B., Hunter, A., Steel, D.: An active contour model for segmenting and measuring retinal vessels. IEEE Trans. Med. Imaging **28**(9), 1488–1497 (2009)
14. Bankhead, P., Scholfield, C.N., McGeown, J.G., Curtis, T.M.: Fast retinal vessel detection and measurement using wavelets and edge location refinement. PLoS ONE **7**(3), e32435 (2012)

15. Chaudhuri, S., Chatterjee, S., Katz, N., Nelson, M., Goldbaum, M.: Detection of blood vessels in retinal images using two-dimensional matched filters. IEEE Trans. Med. Imaging **8**(3), 263–269 (1989)
16. Fraz, M.M., Barman, S.A., Remagnino, P., Hoppe, A., Basit, A., Uyyanonvara, B., Rudnicka, A.R., Owen, C.G.: An approach to localize the retinal blood vessels using bit planes and centerline detection. Comput. Methods Progr. Biomed. **108**(2), 600–616 (2012)
17. Hoover, A., Kouznetsova, V., Goldbaum, M.: Locating blood vessels in retinal images by piecewise threshold probing of a matched filter response. IEEE Trans. Med. Imaging **19**(3), 203–210 (2000)
18. Lesage, D., Angelini, E.D., Bloch, I., Funka-Lea, G.: A review of 3D vessel lumen segmentation techniques: models, features and extraction schemes. Med. Image Anal. **13**(6), 819–845 (2009)
19. Lowell, J., Hunter, A., Steel, D., Basu, A., Ryder, R., Kennedy, R.L.: Measurement of retinal vessel widths from fundus images based on 2-D modeling. IEEE Trans. Med. Imaging **23**(10), 1196–1204 (2004)
20. Martinez-Perez, M.E., Hughes, A.D., Thom, S.A., Bharath, A.A., Parker, K.H.: Segmentation of blood vessels from red-free and fluorescein retinal images. Med. Image Anal. **11**(1), 47–61 (2007)
21. Mendonca, A.M., Campilho, A.: Segmentation of retinal blood vessels by combining the detection of centerlines and morphological reconstruction. IEEE Trans. Med. Imaging **25**(9), 1200–1213 (2006)
22. Soares, J.V., Leandro, J.J., Cesar, R.M., Jelinek, H.F., Cree, M.J.: Retinal vessel segmentation using the 2-D gabor wavelet and supervised classification. IEEE Trans. Med. Imaging **25**(9), 1214–1222 (2006)
23. You, X., Peng, Q., Yuan, Y., Cheung, Y.M., Lei, J.: Segmentation of retinal blood vessels using the radial projection and semi-supervised approach. Pattern Recogn. **44**(10–11), 2314–2324 (2011)
24. Zhang, B., Zhang, L., Zhang, L., Karray, F.: Retinal vessel extraction by matched filter with first-order derivative of Gaussian. Comput. Biol. Med. **40**(4), 438–445 (2010)
25. Carrera, E.V., González, A., Carrera, R.: Automated detection of diabetic retinopathy using SVM. In: 2017 IEEE XXIV International Conference on Electronics, Electrical Engineering and Computing (INTERCON), pp. 1–4. IEEE (2017)
26. Makram-Ebeid, S.: Method and device for automatic segmentation of a digital image using a plurality of morphological opening operation. US Patent 6,047,090, 4 April 2000
27. Chaabane, S.B., Sayadi, M., Fnaiech, F., Brassart, E.: Color image segmentation using automatic thresholding and the fuzzy c-means techniques. In: The 14th IEEE Mediterranean Electrotechnical Conference, MELECON 2008, pp. 857–861. IEEE (2008)
28. Guo, H., Ono, N., Sagayama, S.: A structure-synthesis image inpainting algorithm based on morphological erosion operation. In: Congress on Image and Signal Processing, CISP 2008, vol. 3, pp. 530–535. IEEE (2008)
29. Reza, A.M.: Realization of the contrast limited adaptive histogram equalization (clahe) for real-time image enhancement. J. VLSI Sig. Proc. Syst. Sig. Image Video Technol. **38**(1), 35–44 (2004)
30. Hoover, A., Goldbaum, M.: Locating the optic nerve in a retinal image using the fuzzy convergence of the blood vessels. IEEE Trans. Med. Imaging **22**(8), 951–958 (2003)

Floor Projection Type Serious Game System for Lower Limb Rehabilitation Using Image Processing

Kazuo Hemmi[1(✉)], Yuki Kondo[2], Takuro Tobina[3], and Takeshi Nishimura[4]

[1] Department of Information Systems, University of Nagasaki, Nagasaki, Japan
hemmi@sun.ac.jp
[2] Department of Information and Media Studies, University of Nagasaki, Nagasaki, Japan
knduyuk1226@gmail.com
[3] Department of Nutrition Science, University of Nagasaki, Nagasaki, Japan
tobitaku@sun.ac.jp
[4] Department of Rehabilitation, Miyazaki Hospital, Kuyamacho Isahaya 854-0066, Japan
m-riha@sankoukai.net

Abstract. We described about a rehabilitation system for lower limbs using floor projection type serious game in this paper. In this system, the target is displayed on the floor, and when the user steps on the target, the target disappears and the score is added. Shoot the movement of the user's foot with a TV camera and judge whether or not the target has been stepped by using image processing. The user stands in front of the display area, steps on the target as the target appears. When the target disappears, put the foot back and return to the first posture. By repeating this movement, the user can exercise the lower limbs. The exercise performed in this system is called a Step Exercise and is commonly performed as rehabilitation effective for improving the sense of balance and increasing the muscular strength of the lower limbs. We measured the number that could be erased and the number that could not be erased (the number the system did not respond) when stepping on the target by the foot, and calculated the error rate. The experiment time was 90 s. As a result, the error rate was 4.7%, and the success rate was 95.3%. This was a very high recognition rate. We could judge this system has sufficiently practical accuracy as a floor projection type serious game system for lower limb rehabilitation.

Keywords: Rehabilitation · Serious game · Human computer interaction · Image processing

1 Introduction

According to the "Annual Report on the Aging Society [Summary] FY2017" published by the Cabinet Office Japan [1], the elderly population (over 65 years old) in Japan was 34.59 million people, population ratio was 27.3%. This is the highest figure ever.

P. Boonyopakorn et al. (Eds.): IC2IT 2019, AISC 936, pp. 119–128, 2020.
https://doi.org/10.1007/978-3-030-19861-9_12

According to the report, after this year, the aging of Japan rapidly progresses, and it is expected that the aging problem will become a major issue for Japan in the future. In the current social situation, it is very difficult to suppress the increase in elderly people. However, it is possible to extend the healthy age without care. In Japan it is the most important task to prolong the healthy life of elderly people in the future. If we take thick measures against elderly care prevention and rehabilitation, it is possible to extend the healthy life of the elderly.

Exercise is very important in the elderly care prevention and rehabilitation field. For this purpose, it is important to have a continuous exercise properly. In order to continue exercise, it is desirable to have a system that allows you to enjoy yourself. For that reason, researches on a system capable of enjoying exercise like a game have been actively conducted. This kind of system is called serious game.

Susi et al. [2] outlines the fields of serious games such as military games, government games, educational games, corporate games and medical games. The games of these fields can be categorized into the following two categories. That is, it is training in knowledge field (education field in broad sense) and physical training field (such as rehabilitation). Studies applying serious games to the education field are as follows. Damasceno et al. [3] are conducting research aimed at demonstrating the use of digital games developed for the prevention of drug use, using language and approach for young people. Briscoe et al. [4] propose a serious game which aims to provide better insight and understanding of seaborne trade mechanisms and seafaring practices in the eastern Mediterranean during the Classical and Hellenistic periods. Kobeissi et al. [5] proposes i-Vertex, a hardware framework for concrete serious game on geometric concepts for elementary school students. Andres et al. [6] present the development and validation of a serious game that combines autostereoscopy and Natural User Interfaces for dental learning in higher education.

Studies applying serious games to the rehabilitation are as follows. Funabashi et al. [7] developed a VR serious game (named AGaR) to support the recovery of post-stroke patients. Their game provide a virtual environment that allows the execution of association tasks, where players have to link two images that have complementary meanings. They reported that the game was effective for some participants to improve their upper limb movement. Omelina et al. [8] proposed a specialized configurable architecture for revalidation games, focusing on neuro-muscular rehabilitation. They have been tested with children suffering from cerebral palsy and with their therapists. Through the tests, they observed a significant potential of our approach for building serious games and a positive feedback from both, patients and therapists. Aranha et al. [9] proposed an approach to enable affective adaptation in serious games for motor rehabilitation with physiotherapists' aid. The results of an experiment with physiotherapists show that the system presents a high level of acceptance. Schonauer et al. [10] reported the research about the implementation of a system providing multimodal input, including our own full body motion capture system, a low cost motion capture system (Microsoft Kinect) and biosignal acquisition devices to a game engine. In the research, a workflow has been established, that enables the use of the acquired multimodal data for serious game in a medical environment. A serious game has been implemented, targeting rehabilitation of patients with chronic pain of the lower back and neck. They reported that on clinical change there was a positive trend of decreasing pain intensity score and disability scores and an increase on walking distance after a gaming/training period of 4 weeks. Also, the same research group member [11], presented a full body motion capture (MoCap) system, which, together with a biosignal acquisition device, has been integrated in a game engine. In this study, a workflow has been established that enables the use of acquired skeletal data for serious games in a medical environment. They reported the test results are promising. Ma et al. [12] are studying the development and testing of a serious-game based movement therapy aimed at encouraging stroke patients with upper limb motor disorders to practice physical exercises. Their system contains a Virtual Reality (VR) games. In their report, they mentioned the development and testing of a serious-game based movement therapy aimed at encouraging stroke patients with upper limb motor disorders to practice physical exercises. They reported that the integration of VR simulation with serious games adds richness to the virtual environment which has the potential for improving patient outcome. Hashimoto et al. [13] are conducting research on a system for displaying targets on a 90 cm square force plate. As a result, it showed that the action of stepping on the foot is a moderate exercise load. However, in this system, since the force plate is used, the position of the target can't be freely set.

Our research group has proposed a new human interface by detecting movement of a person using a TV camera, and capable of turning on/off the switch without contact [14–16]. By applying this method, we constructed a floor surface display type rehabilitation system for lower limb in this paper. In this system, the target is projected by the projector, and it is judged whether or not the target has been stepped by using image processing. Therefore, this system has the feature that the display position of the target can be freely set. The user stands in front of the display area, steps on the target as the target appears. When the target disappears, put the foot back and return to the first posture. This exercise is called a Step Exercise and is commonly performed in rehabilitation field for improving the sense of balance and increasing the muscular strength of the lower limbs. The system proposed in this paper aims to perform Step Exercise as a serious game. And, in order to examine the operation accuracy, we conducted a system evaluation experiment. We conducted a total 825 experiments using 19 subjects, we got 95.3% accuracy.

2   Background

We have been studying serious gaming system for upper limb using TV camera so far. In this system, as shown in Fig. 1, the user of this system is shot with a TV camera, and the user's figure and target are projected on the screen. If the image of the hand (or part of the user's body) enters the target area, it is determined that the target has been touched and the target is erased. Judgment when a hand enters the target area is done non-contact by image processing. Therefore, the user does not need to touch the screen directly like a touch screen, and it is only necessary to move the hand over the target in the air while watching the screen. In this system, the target is displayed on the screen with the projector, and the target disappears when touching the target with the body. This system adds points by erasing the targets appearing one after the other with hands, so we can enjoy with the same feeling as a game. By moving the hand, and sequentially erasing the targets, the user can perform the exercise of the upper limb enjoyably with the game feeling. It is possible to increase or decrease the amount of exercise depending on the distance from the TV camera.

In this paper, we describe a method of constructing a lower limb rehabilitation system by using the principle of the upper limb rehabilitation system.

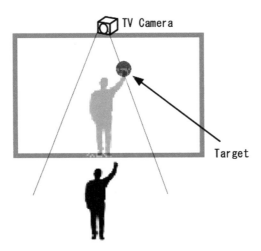

Fig. 1. Rehabilitation system for upper limb

3   System Configuration

In this research, we developed a rehabilitation system for lower limbs using serious game. Figure 2 shows a conceptual diagram of this system. In this system, the target is displayed on the floor, and when the user steps on the target, the target disappears and the score is added. Shoot the movement of the user's foot with a TV camera and judge whether or not the target has been stepped by using image processing. The equipment required for this system is a web camera, projector, PC. The projector is RICHO PJWX 4152N. The TV camera is Logicool C270. All the devices are commercially available products and do not require special equipment.

The user stands in front of the display surface and steps on the target with the foot when the target appears. When the target disappears, put the foot back and return to the first posture. By repeating this movement, the user can exercise the lower limbs. The exercise performed in this system is called a Step Exercise and is commonly performed as rehabilitation effective for improving the sense of balance and increasing the muscular strength of the lower limbs. When performing rehabilitation, if the load is too low, the training effect is insufficient, and if the load is too high, it becomes dangerous. The experimental time was determined to 90 s in consideration of the balance between the effectiveness and danger of training.

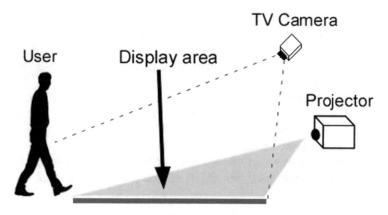

Fig. 2. Concept diagram

The display image of the floor is shown in Fig. 3. As the floor color is often used in dark colors, it is preferable that the display color is bright in order to ensure visibility. Therefore, we adopted a light blue assuming a pond as the background color of the display and a frog as the target. In the display area, the target, the remaining time, the score (the number of times the target has been stepped on), and the foot image of the user shot by TV camera are displayed.

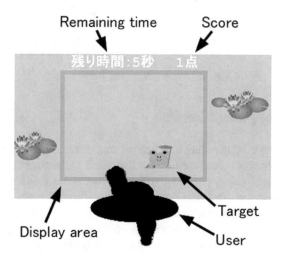

Fig. 3. Floor display image

A block diagram of this system is shown in Fig. 4. This system consists of Touch detection component, Target display component, Counter for points, Timer and Display pattern database. The function of each component is as follows.

- Touch detection component: This component determines from the video image of the TV camera, whether the user has stepped on the target with the foot. When stepping on it outputs a signal "stepped on".
- Counter for points: This counter counts the number of times the user steps on the target with the foot.
- Target display component: This component makes image to be displayed on the floor with the projector by combining the target, the image of the user, the score, and the remaining time of the experiment this component.
- Timer: In this part, the remaining time of the experiment is measured
- Display pattern database: In this database, coordinates for displaying targets are stored and as the experiment progresses, the coordinates of the target are output.

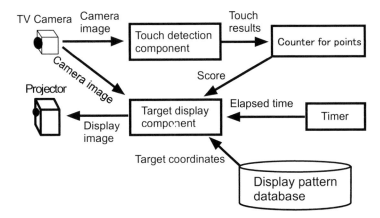

Fig. 4. Block diagram of the system

4 Collision Judgement Method

This system displays images including target and user image on the floor with a projector. Dark colors are often used on the floor, and the surface is not a single color due to patterns, dirt, bruise, etc., so the floor is not quite suitable as the projected surface of the projector. And, this system judges whether or not a target has been stepped (so-called collision judgment) by performing image processing on the video taken by the TV camera. Therefore, the image shot by the TV camera includes the foot image of the user and the image projected by the projector. If the image projected by the projector does not have sufficient brightness, it is dark and is difficult to see by the user, so the image displayed by the projector must be very bright compared to the foot image of the user. Furthermore, the coordinates at which the target is displayed change each time (the target moves). It is very difficult to stably perform a collision judgment between a foot image with very low brightness and a target with high brightness and position change.

Even in this environment, in order to stably perform the collision judgment against the target, in this paper, adopts the collision judgment method using the two-dimensional correlation coefficient. Since the two-dimensional correlation coefficient is normalized with respect to brightness, in principle, a stable result can be obtained against brightness change. Whether the foot has touched the target area (whether or not the target has been stepped on with the foot) is judged by using the two-dimensional correlation coefficient.

From the TV camera, images of 30 frames are sent sequentially to the system every second. Here, we call the currently displayed video frame as the current image, and call the video frame that was displayed immediately before as the previous image.

Two-dimensional correlation coefficient R is defined by Eq. (1). Here, $f(x, y)$ represents the tone value of the pixel at the coordinate (x, y) of the previous image, and $g(x, y)$ represents the tone value of the pixel at the coordinate (x, y) of the current image. $\bar{f}$ represents the average value of the tone values in the previous image, and $\bar{g}$ represents the average value of the tone value in the current image.

$$R = \frac{\sum\limits_{x}^{m}\sum\limits_{y}^{n}(f(x,y) - \bar{f})(g(x,y) - \bar{g})}{\sqrt{\sum\limits_{x}^{m}\sum\limits_{y}^{n}(f(x,y) - \bar{f})^2}\sqrt{\sum\limits_{y}^{n}\sum\limits_{v}^{n}(g(x,y) - \bar{g})^2}} \tag{1}$$

Our system calculates the two-dimensional correlation coefficient between the target area in the current image and the target area in the previous image. Figure 5 shows the concept of the collision judgment method employed in this paper. Figure 5 (a) shows the previous image and the current image when the foot does not enters the target area. In the case of Fig. 5(a), since the previous image and the current image are the same, the two-dimensional correlation coefficient R is one. Figure 5(b) shows the previous image and the current image when the foot enters the target area. In the case of Fig. 5(b), since the previous image and the current image are different, the two-dimensional correlation coefficient R is less than one. We perform a collision judgment using the change in the value of this two-dimensional correlation coefficient R.

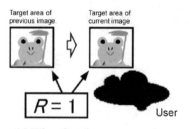

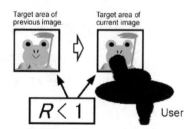

(a) When foot does not enter the target area

(b) When foot entered the target area

Fig. 5. Collision judgment method

5   Evaluation of the System

In order to examine the operation accuracy of the rehabilitation system proposed in this paper, a system evaluation experiment was conducted. In this system, it is important whether the user can accurately erase the target when stepping on the target by the foot. Therefore, we measured the number that could be erased and the number that could not be erased (the number the system did not respond) when stepping on the target by the foot, and calculated the error rate.

The arrangement of the experimental system is shown in Fig. 6. We set a projector (RICHO PJWX 4152 N) at a height of 17 cm from the floor. At this time, the size of the projection area is 1.1 m width and 0.7 m depth. To cover this projection area, we installed a TV camera (Logicool C270) above the projector at a height of 0.75 m from the floor. The position where the subject stands was set 0.1 m in front of the projection area.

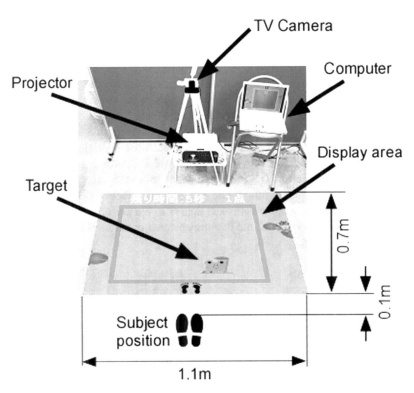

Fig. 6. The arrangement of the experimental system

The total number of subjects was 19, among them 9 men and 10 females. The age of the subjects was 14 in the twenties, 3 in the forties and 2 in the fifties. When the experiment starts, the target is sequentially displayed in the display area, the subject erases the target by stepping on it. The experiment time is 90 s. The number of targets that appeared in 90 s was recorded. Also, we recorded the number of targets that disappeared when stepping on the target (the number the system responded) and the number that did not disappear (the number the system did not respond).

The experimental results are shown in Table 1. Table 1 shows the number of targets that have appeared, the number of times they responded to the movement of the foot, and the number of times they did not responded to the movement of the foot, and the error rate at that time. The total number of trials of 19 subjects was 825. There were 39 trials where the system failed to recognize the foot movement of the subject. The error rate in experiment was 4.7%. This was a success rate of 95.3%, which was a very high recognition rate. From this result, we can judge this system has sufficiently practical accuracy as a floor projection type serious game system for lower limb rehabilitation.

Table 1. Experimental results.

	Total number of times	Number of times that respond to foot movement	Number of times that did not respond to foot movement
Results	825	786 (95.3%)	39 (4.7%)

6 Conclusions

We have described a rehabilitation system for lower limbs using serious game in this paper. In this system, the target is displayed on the floor, and when the user steps on the target, the target disappears and the score is added. Shoot the movement of the user's foot with a TV camera and judge whether or not the target has been stepped by using image processing. The user stands in front of the display area, steps on the target as the target appears. When the target disappears, put the foot back and return to the first posture. By repeating this movement, the user can exercise the lower limbs.

The exercise performed in this system is called a Step Exercise and is commonly performed as rehabilitation effective for improving the sense of balance and increasing the muscular strength of the lower limbs. We measured the number that could be erased and the number that could not be erased (the number the system did not respond) when stepping on the target by the foot, and calculated the error rate. The total number of subjects was 19, among them 9 men and 10 females. The age of the subjects was 14 in the twenties, 3 in the forties and 2 in the fifties. The experiment time was 90 s. As a result, the error rate was 4.7%, and the success rate was 95.3%. This was a very high recognition rate. We could judge this system has sufficiently practical accuracy as a floor projection type serious game system for lower limb rehabilitation.

This time, we conducted a test for healthy subjects, but we need to test for people who need rehabilitation in the rehabilitation department of the hospital next. This will be our future work.

Acknowledgment. This work was supported by JSPS KAKENHI Grant Numbers JP18K11403.

References

1. Cabinet Office Japan: Annual Report on the Aging Society [Summary] FY2017 (2017). http://www8.cao.go.jp/kourei/english/annualreport/2017/pdf/c1-1.pdf
2. Susi, T., Johannesson, M.L., Backlund, P.: Serious games-an overview. Technical report, HS-IKI-TR-07-001, School of Humanities and Informatics, University of Skovde, Sweden (2007)
3. Damasceno, E.F., Nardi, P.A., Silva, A.C.A.: A serious game to support the drug misuse prevention for teenagers students. In: 2017 IEEE Frontiers in Education Conference (FIE) (2017)
4. Briscoe, O.P., Simon, B., Mudur, S.: A serious game for understanding ancient seafaring in the Mediterranean sea. In: 2017 9th International Conference on Virtual Worlds and Games for Serious Applications (VS-Games), pp. 1–7 (2017)
5. Kobeissi, A.H., Sidoti, A., Bellotti, F., Berta, R., Gloria, A.D.: Building a tangible serious game framework for elementary spatial and geometry concepts. In: 2017 IEEE 17th International Conference on Advanced Learning Technologies (ICALT), pp. 173–175 (2017)
6. Andres, D.R., Juan, M.C., Molla, R.: A 3D serious game for dental learning in higher education. In: 2017 IEEE 17th International Conference on Advanced Learning Technologies (ICALT), pp. 111–115 (2017)
7. Funabashi, A.M.M., Aranha, R.V., Silva, T.D.: AGaR: a VR serious game to support the recovery of post-stroke patients. In: 2017 19th Symposium on Virtual and Augmented Reality (SVR), pp. 279–288 (2017)
8. Omelina, L., Jansen, B., Bonnechere, B., Jan, S.V.S., Cornelis, J.: Serious games for physical rehabilitation: designing highly configurable and adaptable games. In: Proceedings of 9th International Conference Disability, Virtual Reality and Associated Technologies, pp. 195–201 (2012)
9. Aranha, R.V., Silva, L.S., Chaim, M.L., Nunes, F.L.S.: Using affective computing to automatically adapt serious games for rehabilitation. In: 2017 IEEE 30th International Symposium on Computer-Based Medical Systems, pp. 55–60 (2017)
10. Schonauer, C., Pintaric, T., Kaufmann, H.: Chronic pain rehabilitation with a serious game using multimodal input. In: 2011 International Conference on Virtual Rehabilitation, 27–29 June 2011 (2011)
11. Schonauer, C., Pintaric, T., Kaufmann, H., Kosterink, S.J., Hutten, M.V.: Full body interaction for serious games in motor rehabilitation. In: Proceedings of the 2nd Augmented Human International Conference, Tokyo, Japan, 12–14 March 2011, Article no. 4 (2011)
12. Ma, M., Bechkoum, K.: Serious games for movement therapy after stroke. In: 2008 IEEE International Conference on Systems, Man and Cybernetics, pp. 1872–1877 (2008)
13. Hashimoto, W., Nakaizumi, F., Inoue, Y., Ohsuga, M.: Interactive multi-modal system for health promotion and preventive care. In: LIFE 2010, Toyonaka, Japan, pp. 87–90 (2010)
14. Hemmi, K.: A spatial operating interfaces and an applications to the interactive digital signage. Technical report, IEICE Technical report, WIT2014-42, pp. 24–25 (2014)
15. Tabata, T., Hemmi, K.: A walking direction detecting method for floor display system. In: The IEEE 9th International Conference on Mobile Ad-hoc and Sensor Networks, Dalian, China, pp. 533–538 (2013)
16. Hemmi, K.: Proposal of image effects for interactive digital signage using image processing by single camera. In: JCTU International Conference "Communication and Changing Society", Bangkok, Thailand, 21 July 2017, pp. 16–25 (2017)

Image Processing Technique for Gender Determination from Medical Microscope Image

Wichan Thumthong[1]($\boxtimes$), Hathaichanok Chompoopuen[2],
and Pita Jaupunphol[3]

[1] Faculty of Computer Science and Information Technology,
Rambhai Barni Rajabhat University, Chanthaburi, Thailand
wichan.t@rbru.ac.th
[2] Faculty of Medicine, Chiang Mai University, Chiang Mai, Thailand
[3] Department of Digital Technology, Phuket Rajabhat University,
Phuket, Thailand

Abstract. This article proposes an image processing technique to calculate the histological variables from microscope image of occipital bones for gender determination. Occipital bones are highlighted based on 13 parameters to find out the relationship with the gender of samplings. The samples are 80 images classified into 46 males and 34 females with ages between 25 and 90 years old. Direct and stepwise discriminant functions are two methods for accuracy evaluation. The experiments show demonstrative results of gender determination accuracy based on both direct and stepwise discriminant functions. While the direct discrimination on 5 parameters represents 97.5% for both gender classification and gender prediction, the stepwise discrimination on 3 parameters is also at 97.5% for both gender classification and gender prediction.

Keywords: Forensic science · Gender determination · Image processing

1 Introduction

Biological identification consists of four basic data types, including gender, age-at-death estimation, height and nationality. Since biological identification plays a crucial role in the individual specification, gender determination has become a vital stage in forensic science [1]. The most reliable non-metric based research considers gross morphological features of the pelvis [2, 3]. Initially, a traditional approach for understanding sexually dimorphic features of the pelvis [2] cranium [4] was dependent upon subjective visual assessment. In Thailand, several gender determination studies have applied the gross morphological analysis to determine sex, such as the lumbar vertebrae [5], metacarpals [6] and talus [7]. Different sex estimation metrics were conducted on various skeletal elements, e.g., scapula [8], sternum and proximal hand phalanges [9].

The disadvantage of gross morphological aging and sexing methods are that they usually require a nearly complete and well preserved skeleton. In many cases,

© Springer Nature Switzerland AG 2020
P. Boonyopakorn et al. (Eds.): IC2IT 2019, AISC 936, pp. 129–137, 2020.
https://doi.org/10.1007/978-3-030-19861-9_13

skeletonremains are unable to be identified by traditional gross morphology features to estimate a biological profile when they are fragmented, poorly preserved, and incomplete. In case where there is no clear indication of individualized traits, fragment bone, anthropologists have turned to bone or dental histology, or the study of microscopic tissues [10].

Nowadays, image processing has been applied in many pathological projects. MATLAB have also become an essential image processing tool in terms of image segmentation [11], cell measurement, cell counting, degraded cell measurement, and brain extraction from MRI images [12]. Medical image processing is one of the most challenging and emerging topics in today research field [13]. In the meantime, biomedical image processing is also considered as the most challenging and upcoming field in the present world [11]. In this research, the image processing technique will be used to calculate the histological parameters from microscope image for automatic gender determination from occipital bones.

2   Research and Methodology

2.1   Graphical Method

The samples consist of 80 images divided into 46 males and 34 females of age between 25 and 90 years with identified age, sex and the cause of death. All the samples were delivered to the Forensic Medicine Department for routine autopsy. The fresh frozen un-embalmed cadavers were received as body donors from the Department of Anatomy for medical surgical training at the Cadaveric Surgical Training Center, Faculty of Medicine, Chiang Mai University, Thailand. The digital images of bones were observed under light microscope at 10x magnification and formatted into JPG with 1600 × 1200 pixels.

2.2   Image Processing Method

The image processing technique and the algorithm source code for analyzing histological parameters are written using MATLAB 2016a, which can be described into 3 stages below.

Pre-processing Stage. The original image of occipital bone is given as the input. The image parameters are separated into four sections with the area of intact osteon and fragmented osteon. While the position of intact osteon is drawn and replaced with red, the fragmented osteon position is replaced with green. Then, the original image background-color is replaced with black color and converted to grayscale.

The brightness levels of the red (R), green (G) and blue (B) components are each represented as 0 to 255 in decimal, or 00000000 to 11111111 in binary. The lightness of the gray is directly proportional to the number representing the brightness levels of the primary colors [13].

Threshold Segmentation Stage. The pixels are divided according to the intensity value. This method is based on the threshold value in which the gray scale image is converted to the binary image. Local methods are deployed to transform the threshold value of each pixel to the local image characteristics for segmentation [11] as shown in Fig. 1.

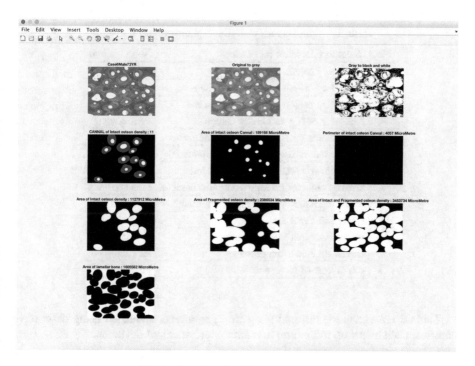

Fig. 1. Results of post-processing stage.

Post-processing Stage. The regionprops is used to measure the image selected region in pixel. The regionprops instruction is used to estimate the scaled area, which is the actual number of pixels in the selected region [14]. The pixel count of the proposed image is dependent upon the distance between the camera and the object when the picture is taken. A reference object is an object with known area required to translate the pixel count area [14].

In this study, the photograph is formatted in a digital JPG image at 1,600 pixels width and the object of femur bone width is 1406.36 μm. Therefore, the pixel value calculation is based on the following.

$$1 \; pixel \; value = width \; object / width \; image; \; 1 \; pixel \; value = 0.879 \; \mu m^2; \quad (1)$$

Please note that the area of parameters (μm²) will be X * 1 pixel if the pixel count area is X pixels. Post-processing stage from the original image is represented in Fig. 1.

2.3   Histological Parameters

Thirteen histomorphometric parameters with their abbreviations are clarified in Table 1.

Table 1. Abbreviations of histomorphometric parameters.

No.	Histomorphometric parameters	Abbreviations
1	Secondary osteon area	On.Ar
2	Secondary osteon maximum diameter	On.max
3	Secondary osteon minimum diameter	On.min
4	Perimeter of big secondary osteon	Pm.big.on
5	Big secondary osteon maximum diameter	Big.on.max
6	Big secondary osteon minimum diameter	Big.on.min
7	Perimeter of small secondary osteon	Pm.small.on
8	Small secondary osteon maximum diameter	Small.on.max
9	Small secondary osteon minimum diameter	Small.on.min
10	Haversian canal area	Hc.Ar
11	Haversian canal max diameter	Hc.max
12	Haversian canal min diameter	Hc.min
13	peTrimeter of haversian canal	Pm.Hc

Table 2 represents the histomorphometric parameters calculated using descriptive statistics, such as mean, minimum, maximum, and standard deviation.

Table 2. Occipital parameters for the pooled genders.

Parameters	Mean	S.D.	Min	Max
1. On.Ar[a] (μm^2)	4.883	0.086	4.50	5.00
2. On.max[a] (μm)	4.601	0.093	4.44	4.82
3. On. min[a] (μm)	4.387	0.168	3.79	4.91
4. Pm.big.on (μm)	655.013	108.481	503.00	898.00
5. Big.on.max (μm)	273.850	54.132	201.00	421.00
6. Big.on.min (μm)	177.088	30.779	74.00	260.00
7. Pm.small.on (μm)	98.225	34.991	30.00	229.00
8. Small.on.max (μm)	200.388	48.546	93.00	375.00
9. Small.on.min (μm)	144.350	28.412	75.00	242.00
10. Hc.Ar [a] (μm^2)	3.612	0.140	3.33	3.84
11. Hc.max[a] (μm)	3.404	0.125	3.09	3.66
12. Hc.min[a] (μm)	3.291	0.114	3.02	3.47
13. Pm.Hc (μm)	142.363	36.718	63.00	262.00

[a]Log 10, square micrometers (μm^2), μm = micrometers

2.4   Statistical Analysis

Descriptive statistics used to summarize and calculate indices of both males and females are illustrated in Table 3. The results show that 6 parameters are significantly different (P < 0.05) between males and females. However, 7 parameters, including On. Min, Big.on.max, Big.on.min, Pm.small.on, Small.on.max, Small.on.min, and Pm.Hc, are not considered. The results show that there is no statistical significance (P > 0.05) between males and females.

Table 3. Comparation of occipital histological parameters between female and male.

Parameters	Female (n = 35)		Male (n = 45)		t-value	p-value
	Mean	S.D.	Mean	S.D.		
1. On.Ar^a (μm^2)	4.946	0.043	4.832	0.077	−8.409	0.000*
2. On.max^a (μm)	4.663	0.081	4.550	0.068	−6.810	0.000*
3. On. min^a (μm)	4.370	0.175	4.402	0.163	0.841	0.403^{NS}
4. Pm.big.on (μm)	746.028	84.082	580.545	57.066	−10.064	0.000*
5. Big.on.max (μm)	260.917	52.565	284.432	53.664	1.968	0.053^{NS}
6. Big.on.min (μm)	177.806	32.683	176.500	29.499	−0.188	0.852^{NS}
7. Pm.small.on (μm)	92.528	33.705	102.886	35.713	1.324	0.190^{NS}
8. Small.on.max (μm)	195.25	50.649	204.59	46.921	0.855	0.395^{NS}
9. Small.on.min (μm)	139.75	29.657	148.11	27.111	1.316	0.192^{NS}
10. Hc.Ar^a (μm^2)	3.487	0.090	3.714	0.074	12.363	0.000*
11. Hc.max^a (μm)	3.465	0.086	3.465	0.086	5.721	0.000*
12. Hc.min^a (μm)	3.191	0.087	3.373	0.049	11.185	0.000*
13. Pm.Hc (μm)	136.000	38.598	147.568	34.679	1.410	0.162^{NS}

n: Number of samples, ^{a}Log 10, *Significant (P < 0.05),
^{NS}Insignificant (P > 0.05)

Direct and stepwise discriminant functions are two evaluation criteria for gender determination accuracy. The discriminant analysis shows that the gender determination accuracy using the discriminant function in the direct method and the classification function formulae for all variables are clarified in Table 4. The results indicate that parameter 1 is a prominent indicator for genders determination. Please noted that Hc.Ar alone is capable of classifying the gender at 98.8% for both original and cross validated groups. In this case, the accuracy obtained by using direct variable, especially the Hc. Ar variable, can be used for gender determination from histomorphometry. Table 4 illustrates the moderate accuracy of gender classification and prediction observed from 5 variables, followed by On.Ar, On.Max, Pm.big.on, Hc.Max and Hc.min, respectively. For the stepwise discriminant function analysis, however, there are three 3 selected variables, including On.Ar, Pm.big.on and Hc.min, from the total of 5 (also see Table 4).

Table 4. Direct univariate discriminant function analysis.

Parameters	Discriminant function equation	Centroids		Average accuracy (%)	
		M	F	O	C
1. On.Ar*	(7.710 × On.Ar) + (−37.020)	−0.377	0.510	87.5	87.5
2. On.Max*	(0.008 × Pm.big.on) + (−4.954)	0.217	−0.293	80.0	80.0
3. Pm.big.on*	(0.021 × Big.on.max) + (−5.648)	0.444	−0.601	85.0	85.0
4. Hc.Ar	(0.21 × Small.on.max) + (−4.603)	−0.220	0.297	98.8	98.8
5. Hc.max*	(8.558 × Hc.Ar) + (−34.271)	1.047	−1.416	80.0	80.0
6. Hc.min*	(19.129 × Hc.max) + (−66.974)	1.399	−1.893	58.8	58.8

* = observed parameters
O = original group correctly classified, C = cross-validated group correctly

3 Results and Discussion

3.1 Results

This study performed 2 methods, as follows.

The results of direct discriminant analysis in Table 5 based on 5 variables, such as On.Ar, On.max, Pm.big.on, Hc.max and Hc.min, show moderate correlations. The discriminant score is calculated by multiplying the unstandardized coefficient with each particular measurement, summarising them, and adding the constant below.

$$
\begin{aligned}
Discriminant\,score =& (6.775 \times On.Ar) + (0.758 \times On.Ar.\mathrm{max}) \\
& + (0.009 \times Pm.big.on) + (-1.286 \times Hc.\mathrm{max}) \\
& + (-8.636 \times Hc.\mathrm{min}) + (-9.750)
\end{aligned}
\tag{2}
$$

Table 5 also shows 3 selected variables from 5 variables in discriminant function, including On.Ar, Pm.big.on, and Hc.min. The discriminant score was calculated as:

$$
\begin{aligned}
Discriminant\,score =& (6.926 \times On.Ar) + (0.009 \times Pm.big.on) \\
& + (-9.476 \times Hc.\mathrm{min}) + (-8.823)
\end{aligned}
\tag{3}
$$

After the discriminant score has been calculated from direct and stepwise discriminant functions, the score is compared with the specified criteria: (1) 0 (a halfway between the female and male centroids); (2) greater than 0 (male); and (3) less than 0 (female).

An eigenvalue indicates the proportion of variance calculated from between-groups sums of squares divided by within-groups sums of squares. A large eigenvalue is associated with a strong function. Based on the eigenvalue in Table 5, each of the eigenvalue of direct and stepwise methods are 4.309 and 4.230 respectively. The eigenvalue must not be less than 1 to confirm the discriminant function accuracy.

Table 5. Coefficients for the stepwise and direct discriminant function analysis.

Parameters	Unstandardized coefficient	Eigen value	Canonical correlation	Wilks' lambda	Centroids	Sectioning point
Direct discriminant function analysis						
1. On.Ar	6.775	4.309	0.901	0.188	M = −1.854	0
2. On.max	0.758				F = 2.266	
3. Pm.big.on	0.009					
4. Hc.max	−1.286					
5. Hc.min	−8.636					
Constant	−9.750					
Stepwise discriminant function analysis						
1.On.Ar	6.926	4.230	0.899	0.191	M = −1.837	0
2. Pm.big.on	0.009				F = 2.245	
3. Hc.min	−9.476					
Constant	−8.823					

Please also note that the greater different values between the groups will imply higher discriminative possibility.

The canonical relation is a correlation between the discriminant scores and the dependent variable levels. A higher correlation indicates a better discriminative function. Canonical correlation is the value representing the relation between independent variables and dependent variables. The canonical correlation value from 0.8 to 1.0 is considered high. Consequently, the analyzed values of this study are 0.901 and 0.899, which are very high.

Wilks' lambda is a measure of how well each function separates cases into groups. Smaller values of Wilks' lambda indicate greater discriminatory ability of the function and that group means appear to differ. Table 4 show the results of direct and stepwise methods, which are 0.188 and 0.191 respectively. This means that the selected independent variables in the discriminant function of two methods are divided into groups. Furthermore, the value of Wilks' Lambda, which has statistical significance (P-value < 0.005), indicates that the group means are varied.

3.2 Accuracy of Classification

The accuracy results for gender determination of the discriminant function derived from 6 variables in the direct method are illustrated in Table 6. The accuracy results for both classification and prediction are 97.5%. In the meantime, the results of those in the stepwise method derived from 3 variables for both classification and prediction are also 97.5%.

Table 6. Classification and prediction accuracy using direct and stepwise discriminant models.

Parameters	Classification accuracy			Prediction accuracy		
	Male	Female	Total	Male	Female	Total
Direct	100.0	94.4	97.5%	100.0	94.4	97.5%
Stepwise	100.0	94.4	97.5%	100.0	94.4	97.5%

3.3 Discussion

The accuracy results for gender determination based on the discriminant function derived from 5 variables in the direct method are 97.5% for both classification and prediction. When the stepwise is applied on 3 variables, the promising results were also represented in similar manner to the direct method. The accuracy at 97.5% for classification and prediction has been witnessed. In this case, the techniques used for this study can be considered as very effective and suitable for forensic science since the accuracy of gender classification and prediction can be observed at 97.5%.

4   Conclusions

An effective approach for gender determination from medical microscope image using image processing and its methodological steps have been proposed in this article. While direct and stepwise discriminant functions are two key approaches underlying the determination accuracy criteria, the experimental results on both functions are demonstrative. However, further experiments on this proposed technique must continue on different medical microscope images to yield the results validity.

References

1. Christensen, A.M., Passalacqua, N.V., Bartelink, E.J.: Forensic Anthropology: Current Methods and Practice, 1st edn. Academic Press (2014)
2. Phenice, T.W.: A newly developed visual method of sexing the Os pubis. Am. J. Phys. Anthropol. **30**(2), 297–301 (1969)
3. Bruzek, J.: A method for visual determination of sex, using the human hip bone. Am. J. Phys. Anthropol. **117**(2), 157–168 (2002)
4. Buikstra, J.E., Ubelaker, D.H.: Standards for data collection from human skeletal remain. In: Proceedings of a seminar at the Field Museum of Natural History, Arkansas Archeological Survey, 12154th edition (1994)
5. Ostrofsky, K., Churchill, S.E.: Sex determination by discriminant function analysis of lumbar vertebrae. J. Forensic Sci. **60**(1), 21–28 (2014)
6. Khanpetcha, P., Prasitwattansereeb, S., Casec, D.T., Mahakkanukrauh, P.: Determination of sex from the metacarpals in a Thai population. Forensic Sci. Int. **217**(1–3), 229.e1-8 (2011)
7. Mahakkanukrauha, P., Praneatpolgranga, S., Ruengdita, S., Singsuwana, P., Duangtob, P., Case, D.T.: Sex estimation from the talus in a Thai population. Forensic Sci. Int. **240**, 152-e1 (2014)

8. Dabbs, G., Moore-Jansen, P.H.: A method for estimating sex using metric analysis of the scapul. J. Forensic Sci. **55**(1), 149–152 (2009)

9. Mahakkanukrauh, P., Prasitwattanseree, S., Casec, D.T.: Determination of sex from the proximal hand phalanges in a Thai population. Forensic Sci. Int. **226**(1–3), 208–215 (2013)

10. Crowder, C., Stout, S.: Bone Histology An Anthropological Perspective, 1st edn. CRC Press (2011)

11. Kumar, D.D., Vandhana, S., Priya, K.S., Subashini, S.J.: Brain tumour image segmentation using MATLAB. Int. J. Innov. Res. Sci. Technol. **1**(12), 447–451 (2015)

12. Patil, R.C., Bhalchandra, A.S.: Brain tumour extraction from MRI images using MATLAB. Int. J. Electron. Commun. Soft Comput. Sci. Eng. **2**(1), 1–4 (2015)

13. Nithyanandam, G.: Extraction of cancer cells from MRI prostate image using MATLAB. Int. J. Eng. Sci. Innov. Technol. **1**(1), 27–35 (2012)

14. Patil, S.B., Bodhe, S.K.: Betal leaf area measurement using image processing. Int. J. Comput. Sci. Eng. **3**(7), 2656–2660 (2011)

Accelerate the Detection Frame Rate of YOLO Object Detection Algorithm

Wattanapong Kurdthongmee[(✉)]

School of Engineering and Resources Management, Walailak University,
222 Thaibury, Tha-sa-la, Nakhon-si-thammarat 80161, Thailand
kwattana@wu.ac.th

Abstract. YOLO (You-Only-Look-Once) is by far the well-known Deep Neural Networks (DNNs) object detection algorithm with real-time performance on a computer with GPUs. Conceptually, YOLO divides the input image of size $W \times W$ into non-overlapping square cells with the final feature of size $S \times S$; i.e. $(416 \times 416) \rightarrow (13 \times 13)$. Each cell is responsible for predicting a single object whose centre falls into it. In this paper, we propose the algorithm that makes use of our observation mapping relationship which states that while the sizes of square cells are changed from layer to layer, their indices are preserved. The algorithm operates by locating a region of change in an input image and identifies the indices of square cells that cover the region. Only the members of the input features within these cells in all layers along the network are required to be operated. When the algorithm is employed along with the spatio-temporal property within video frames, it is capable of attaining the best relative detection of 1.47 (about 7 fps) with 90% correctness. These are benchmarked with the ordinary YOLO object detection on a personal computer: Intel Core i7 CPU at 3.5 GHz with 16 GB of memory and without any sophisticate GPUs, on the Tiny-YOLO network.

Keywords: YOLO object detection · Spatio-temporal property ·
Deep neural networks

1 Introduction

An object detection is a higher level application of DNNs (Deep Neural Networks) above an object classification. While an object classification gives only the prediction probability of an object if it exists within an input image, an object detection is capable to locate the region that covers an object. The output from an object detection usually contains the bounding box around an object along with its prediction probability. YOLO (You-Only-Look-Once) [1,2] is by far the most famous object detection algorithm with real-time performance on a computer platform with a sophisticate GPU. The demands of this platform; in terms of size, cost and power consumption, prevents YOLO to be usable in some applications; i.e. a smart CCTV and an IoT node with object detection capability, where cost, size and power consumption are sensitive.

© Springer Nature Switzerland AG 2020
P. Boonyopakorn et al. (Eds.): IC2IT 2019, AISC 936, pp. 138–147, 2020.
https://doi.org/10.1007/978-3-030-19861-9_14

The high computational intensive of YOLO is within its convolution layers which basically perform the matrix multiplication between an input feature and a set of filters. Efficiency improvements of DNNs at the algorithmic level can be accomplished by focusing on a convolution which can be implemented by three different approaches [3]: im2col algorithm, Winograd algorithm or FFT (Fast Fourier Transform). They can also be leveraged by reducing the multiplicand of the matrix multiplication. Typically, video frames seem to vary smoothly and slightly. This is very obvious if video is captured from a static camera setting. By this way, the spatio-temporal property between frames can be exploited to reduce the size of multiplicand and, in turn, accelerate processing a convolution. In this paper, we propose the algorithm that relies on using the spatio-temporal property in addition to the mapping relationship between the regions of features in different layers of YOLO to accelerate the overall object detection.

This paper is organised as follows. The literatures are surveyed in Sect. 2. The proposed algorithm is detailed in Sect. 3. The implementation and experimentations are described in Sect. 4. Finally, the paper is concluded in Sect. 5

2   Literature Surveys

The im2col based algorithms are mostly accelerated by running the convolution on a sophisticate GPU [3] or special hardwares [4,5]. These are done without taking the spatio-temporal property into account. As far as the frame similarity is concerned, there are two publications that exploit this benefit. The CBinfer presented by [6] relies on detection of a set of modified input features on a member by member basis between consecutive frames. A modified input feature at any location is defined as the difference between the members of input features with a value greater than some threshold. The modified indices extraction is then performed in order to create the multiplicand. Most of the proposed operations attempt to make it possible to parallel process by a GPU. It is claimed that it can reach the average speed-up of 8.6 over the cuDNN using an optimised GPU implementation. Similarly, [7] presented the DeepCache concept that exploits the redundancy computation from the previous frame to reduce computation required in a current frame. The approach searches for similar regions in a frame under consideration by use of a region matcher which is inspired by video compression. The implementation of the matcher is not detailed. The results from the region matcher are used to avoid performing computation. Because they contain some classification or detection knowledge that can be used to find the final classification or detection result. It is claimed that the DeepCache saves the inference execution time by 18% on average with the maximum of 47%.

In the next section, we propose the algorithm to accelerate the YOLO object detection that attempts to reduce the size of members of an input feature to be used for convolution. It differs from the previously proposed ones on the point that it relies on finding the regions of difference within the input features between consecutive frames on a frame by frame basis. This helps reducing the considerable amount of time to compare the input features at a layer level.

3   The Proposed Algorithm to Accelerate the YOLO Object Detection

In this section, we firstly give a brief background of the YOLO object detection algorithm. The relationship between the members of features in some layers are highlighted. Then, we extend the relationship to cover all the layers of the YOLO network in order to accelerate their internal processes.

3.1   Background of the YOLO Object Detection Algorithm

YOLO is in the class of fully convolution network (FCN). This means that the network composes of convolution layers without any fully-connected layers. **It operates by partitioning the input image into non-overlapping square cells. Each cell covers the $(K \times K)$ members on the input image which is the first layer of the network. It is responsible for predicting a distribution over class labels as well as a bounding box for the object whose centre falls into it. The whole detection process gives rise to the final tensor of size** $S \times S \times (B \times 5 + C)$. Suppose the dimension of an input image is $W \times W \times 3$; a square of W width and height with 3 colour channels, followings are the details of the final tensor:

- $S = \frac{W}{K}$. The $(K \times K)$ members of an input image are internally processed by all layers of the network; i.e. convolution and pooling. The final result is stored in a single grid member of the final tensor whose relative location is preserved. This means that the cell covering the $(K \times K)$ members of an input image at the (r, c) location is also mapped to the (r, c) location of the final feature. It can be summarised that the whole detection process reduces the cell members from $(K \times K)$ to (1×1).
- B is a bounding box prediction. It has 5 components which are $(x, y, w, \text{h}, \text{confidence})$. The (x, y) coordinates represent the centre of the box, relative to the cell location and normalised to fall between 0 and 1. The (w, h) box dimensions are also normalised to $[0, 1]$, relative to the image size.
- C is the class probability.

In the next section, we will draw the mapping relationship between cell indices and cell members from the internal layers of YOLO.

3.2   Our Proposed Mapping Relationship

Before going further into details, let us give the definition of some terms that are going to be used for the rest of this paper:

- **A cell** is a region within the feature resulting from partitioning the feature into non-overlapping square regions.
- **Cell members** are the $(K \times K)$ members of the feature within the cell.
- **A cell size** is the total number of cell members within the cell.
- **A cell index** is the location (r, c) of the cell within its set of cells.

As a proof of concept, our research targets on using the downsized version of YOLO; the Tiny-YOLO, with only 16 layers. Table 1 details all layers of the Tiny-YOLO. It is noticeable that the Tiny-YOLO partitions the input image into $(S \times S) = (13 \times 13)$ cells. Each internal pooling layer is responsible for downsizing the input feature by a factor of (2×2) with; once again, region index preservation. This is very importance relationship which is employed to accelerate the object detection in our proposed algorithm.

Table 1. The Tiny-YOLO network: C: Convolution, P: Pooling, and OD: Object Detection. W is the feature width and height. D is the number channels.

Layer	0	1	2	3	4	5	6	7	8	9	10	11	12	13	14	15
Operator	C	P	C	P	C	P	C	P	C	P	C	P	C	C	C	OD
Dimension	W	416	416	208	208	104	104	52	52	26	26	13	13	13	13	13
	D	3		16	16	32	32	64	64	128	128	256	256	512	512	1024	1024

We have already given the relationship between the input image and the final tensor: $(W \times W) \to (S \times S)$. At this point, let us detail the relationships of all the in-between layers. At the layer i^{th}, the cell members $(K_i \times K_i)$ within a cell index (r_i, c_i) are processed to produce the cell members $(K_{i+1} \times K_{i+1})$ as its output. It is reasonable to state that the cell members $(K_{i+1} \times K_{i+1})$ is also within a cell index (r_i, c_i) of the layer $(i + 1)^{th}$. It is noted that we avoid considering the number of channel of a feature as it does not affect by this partitioning. With respect to Table 1, the width and height dimensions of the 1st and final layers are (416×416) and (13×13), respectively. This means that a cell of the 1st layer has cell members of $(\frac{416}{13} \times \frac{416}{13}) = (32 \times 32)$. After performing convolution to a cell of this layer, the output cells with equal size of (32×32) whose number of channels is 16 are obtained. Now, consider the 2nd layer whose pooling operation is operated. It gives rise to half width and height dimensions of its input. It is equivalent to say that after processing the 2nd layer the cell members are reduced to (16×16) while the total number of cells is preserved at (13×13). The cell index (r_1, c_1) has half cell size after processing the cell members within the cell index (r_0, c_0) while its cell index is equal to (r_0, c_0). The same relationship between cell sizes and cell indices are repeated up to the layer 12th where the cell size is constant and equal to (1×1). At this point, let us make the mapping relationship between cells in difference layers of YOLO. **The mapping relationship states that while the cell sizes are changed from layer to layer, their indices are preserved.**

Having known the mapping relationship between different layers of the network, it can now be further employed to accelerate the object detection. This is done by exploitation the spatio-temporal property of video frames. **In this way, if a region of change in a frame; the first input feature to the network, can be located and the cells that cover the region can be identified, only these cells and the cells with the same indices in the next layers**

Input;
$\overline{F_{t-1}, F_t}$: The previous and current frame of input image, respectively.;
Output;
$\overline{F_t^*}$: The current frame of input image with detected objects;
```
/* Find the differences between the current and previous frames.        */
```
$\Delta(r, c, d) \leftarrow T$ if $|F_{t-1}(r, c, d) - F_t(r, c, d)| > \tau_{diff}$;
```
/* Initialise the M array from each (32,32) members of Δ(r,c).         */
```
for $i \leftarrow 0$ **to** *12* **do**
$\quad$ **for** $j \leftarrow 0$ **to** *12* **do**
$\quad\quad (startR, startC) \leftarrow (i \times 32, j \times 32)$;
$\quad\quad (endR, endC) \leftarrow (startR + 32, startC + 32)$;
$\quad\quad$ **if** *Any of* $\Delta((startR, .., endR), (startC, .., endC), (0, .., 2)) == T$ **then**
$\quad\quad\quad |\quad M(i, j) \leftarrow$ 'T';
$\quad\quad$ **else**
$\quad\quad\quad |\quad M(i, j) \leftarrow$ 'F';
$\quad\quad$ **end**
$\quad$ **end**
end
```
/* Group the scattered members of M into regions array M*.            */
/* Each M* member has the following coordinate: (L,T,R,B).            */
```
$M^* \leftarrow Group(M)$;
```
/* Perform object detection.                                          */
```
for *layer* $\leftarrow 0$ **to** *15* **do**
```
   /* Convolution layer.                                              */
```
$\quad$ **if** *Convolution* **then**
$\quad\quad |$ See Algorithm 2
```
   /* Max pooling or region: Using the ordinary operations.           */
```
$\quad$ **else if** *Max pooling Or Region* **then**
$\quad\quad |$ MaxPool or Region;
$\quad$ **end**
end

Algorithm 1. The proposed YOLO object detection algorithm.

along the network are required to recalculate. It is highlighted here that the mapping relationship states that it is not necessary to waste time making comparison of features; between the frame $t + 1$ and t, at the layer level.

In the next section, the proposed algorithm to accelerate YOLO object detection by employing the proposed mapping relationship and the spatio-temporal property is detailed. It is focused but not limited to the Tiny-YOLO network.

3.3   Our Proposed Algorithm

At this point, let us explain the details of our proposed algorithm; Algorithm 1, on a frame by frame basis. All network parameters and operators are based on the Tiny-YOLO network detailed in Table 1. Suppose the previous frame of input image F_{t-1} is explicitly stored, the algorithm first finds the pixel differences between the current and previous frames of input image; F_{t-1} and F_t. The pixels in both frames differ if their absolute of subtraction result in all channels is greater than the threshold τ_{diff}. The comparison results in the Δ matrix are, then, assigned to the M array whose size is $(13, 13)$. A member at (i, j) of the M array is mapped to $(32, 32)$ members of the Δ matrix. Only a single member within $(32, 32)$ members of the Δ matrix changes the state of $M(i, j)$ from 'F' to

'T'. In reality, the M array can be used right away by all layers of the network to selectively prepare their members of input to operate; all members of their input feature which are mapped by the 'T' members of the M array. It is, however, more efficient to group the members of M array from scattered cells into the region array M^* whose format is: (left, top, right and bottom). For example, the $M^* = \{(0,0,1,1),(9,9,11,11)\}$ is produced from the M array whose following indices are 'T': $\{(0, 0), (0, 1), (9, 9), (9, 10), (9, 11), (9, 12), (1, 0), (1, 1), (10, 9), (10, 10), (10, 11), (10, 12), (11, 9), (11, 10), (11, 11), (11, 12)\}$.

After the M^* has been created, the algorithm performs object detection on a layer by layer basis. For the convolution layer, Algorithm 2 is invoked. Within the algorithm, first of all it computes the number of cell members K_L which depends on the layer index L; i.e. $L = 1, K_L = \frac{416}{13} = 32$ and $L = 3, K_L = \frac{208}{13} = 16$. Next, the algorithm visits each member of the region array M^*. For the one whose index is m, its members which keeps the (left, top, right, bottom) coordinates of region are retrieved. These coordinates are the region with respect to 13×13 cells where each cell covers 32×32 members. They are required to convert to the start index (L, T) and end index (R, B) with respect to the input feature A by multiplying them with the number of cell members K_L. The (L, T) and (R, B) are then used to selectively retrieved the members from the input feature A. These are assigned to S and, then, rearranged into column matrix I_S; via Im2Col function, in order to accelerate the convolution. Finally, the matrix I_S is used as a multiplicand to perform convolution operation with the weight W of the layer to produce the D matrix as an output.

There are two optional operations within the convolution layer after the convolution, these are the batch normalisation and the activation. These operations are required to process on the D matrix prior to update the output feature C in the final stage of the algorithm. It is noted that some members of the output feature C are preserved by our algorithm; i.e. the members which are out of the frames difference area. They are inherited from the previous frame since the members of input feature they are mapped to do not change as the differences are less than τ_{diff}. The algorithm only updates the members of the output feature C within the start index (L, T) and end index (R, B).

Our proposed algorithm keeps using the ordinary pooling and detection algorithms to process the feature within the pooling and region layers, respectively.

3.4   Summary of the Relationships Between Layers of the Tiny-YOLO

At this point, let us summarise the relationships between difference areas within the input image, the M array and the modified members within some intermediate and final layers of the Tiny-YOLO. This is illustrated in Fig. 1. Let us start from the left of the figure which shows the video frame with a walking person. The detected regions of difference are bound by the blue rectangle. It is mapped to the region array M^* whose size is (13×13). It produces a single region within M^* with the coverage of $(3, 3, 5, 6)$. These are the information that is going to

Input;
_____
A : The input feature;
M^* : The array that keep states of 'T' or 'F' for each $(32, 32)$ members;
W : The weight matrix;
L : The current layer index;
Input/Output;
_____
C : The convolution matrix;

$K_L \leftarrow$ The number of cell members;
for $m \leftarrow 0$ **to** $|M^*|$ **do**
    if M_m^* *state is 'T'* **then**
        /* Convert the coordinate of M^* to the indices of A.   */
        $(T, L) \leftarrow (M_{m,T}^* \times K_L, M_{m,L}^* \times K_L)$;
        $(B, R) \leftarrow (M_{m,B}^* \times K_L, M_{m,L}^* \times K_L)$;
        /* Retrieve the members of A within the region (T, L, B, R).   */
        /* NB: All channels are required.   */
        $S \leftarrow A_{(T,L),\ldots,(R,B)}$;
        $I_S \leftarrow Im2Col(S)$;
        $D \leftarrow I_S \circledast W$;
        /* Perform batch normalization/activation to D if needed.   */
        ......;
        /* Partially update C with currently processed members.   */
        /* NB: Keep all unchanged regions.   */
        $C_{(T,L),\ldots,(R,B)} \leftarrow D$
    end
end

Algorithm 2. The proposed convolution algorithm.

be used by all layers of the Tiny-YOLO network. Let us start from the 1st convolution. The region of M^* is converted to the region within the input feature of the 1st convolution: $(96, 96, 160, 192)$. Only all the members within this region are required to use to perform convolution. The members in this region are also modified in the output feature of this layer. The 2nd layer whose operation is the pooling only process the similar region to the 1st layer. Its output downsizes the input feature by a factor of (2×2). The 3rd is similar to the 1st layer, it performs convolution to the region indicated by the region array M^* which is mapped to the region $(48, 48, 90, 106)$ within its input feature. All these similar steps apply to the rest layers until the 15th layer.

In the next section, the implementation and experimentations of our proposed algorithm are detailed.

4   Implementation and Experimentations

We implemented the proposed algorithm by use of Python with some C-extension. It was experimented on a personal computer with the following specifications: MacBook Pro with Intel Core i7 running at 3.5 GHz, 16 GB memory, Intel Iris Plus Graphics 650 and MacOS High Sierra operating system. All the execution times were measured by use of Python's **time** module.

The experimentations were performed on the video clip https://youtu.be/ 9Z8HH_dEaWw which is modified from the original one available from: https:// www.youtube.com/watch?v=Nqv-6z7QSqE. The modified video clip consists of two separate almost equal duration scenes of walking persons. Both scenes were

$\Delta = \|F_{t-1} - F_t\| > \tau_{diff}$	M^*	Tiny-YOLO layer/operation			
		1: Convolution	2: Maxpool	3: Convolution	15: Region
416×416	13×13 (1, 1)	$A_W \times A_H$ = 416×416 $K_L \times K_L$ = (32, 32)	208×208	208×208 (16, 16)	13×13 (1, 1)
	M_m = {{3, 3, 5, 6}}	(L, T, R, B) = (96, 96, 160, 192)	-	(48, 48, 90, 106)	(3, 3, 5, 6)

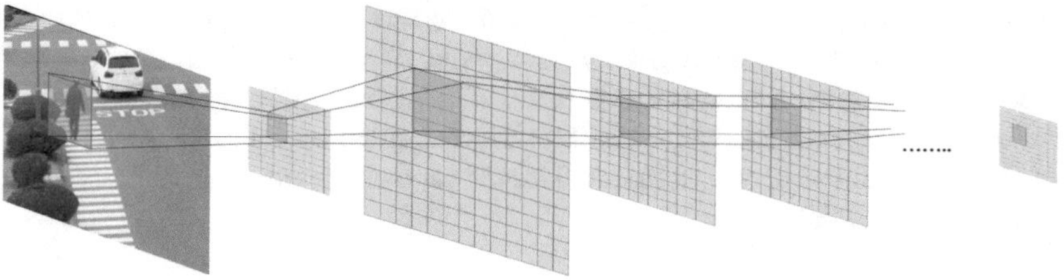

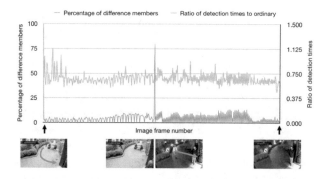

Fig. 1. The relationships between difference areas within the input image, the M array and the modified members within some intermediate and final layers of the Tiny-YOLO.

captured by a static camera setting. Figure 2 at the bottom shows some frames of the video clip. The blue lines represent the track of walking persons within the clip. It is our intension to rely on this kind of video clip since the acceleration ratio of our proposed algorithm relies heavily on the degree of spatio-temporal property. All experimentations were performed to process 200 frames of the video clip.

Fig. 2. The relationships between the percentage of difference members; $\|F_{t-1} - F_t\| > \tau_{diff}$ when $\tau_{diff} = 0.1875$, in blue and the ratio of detection times in green. Some input image frames are shown under the graph.

The first experiment was aimed to study the relationship between the percentage of difference members within the input frames; $\|F_{t-1} - F_t\| > \tau_{diff}$ to the ratio of detection times between our proposed algorithm and the ordinary YOLO detection. The experiment results; $\tau_{diff} = 0.1875$, are shown in Fig. 2. The percentage of difference members and the ratio of detection time improvements are represented in blue and green, respectively. It is obvious that the ordinary YOLO object detection slightly outperforms our proposed algorithm at the beginning of both scenes of the video frame (see the images under the graph). This results from the fact that almost all members of the Δ matrix are true at these frames which means that almost all pixels in the F_t^{th} frames differ from the F_{t-1}^{th} frames. These give rise to a single member of the region array M^* with the coverage of $(0, 0, 12, 12)$ which signals all layers of the network to take all members to perform their operations. After the first frames onward, our proposed algorithm outperforms the ordinary one with the average ratio of detection time improvement of 0.68. The detection times are faster as a result of the reduction of the feature members. The average of the percentage of difference members is only 4.36%.

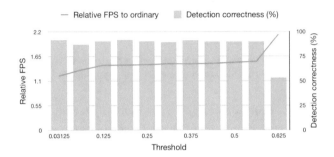

Fig. 3. The relationships between the relative detection frame rates and the values of difference threshold τ_{diff} between our proposed algorithm and the ordinary one.

In the second experiment, we studied the impact of the difference threshold τ_{diff} to the detection correctness and the relative detection frame per second between our proposed algorithm and the ordinary YOLO object detection. The τ_{diff} was varied from 0.03125 to 0.625 with the increment of 0.03125. The experiment results are shown in Fig. 3. The detection correctness is shown as a bar graph while the relative detection frame per second is plotted as a line graph. It is obvious that the maximum detection correctness occurs at the minimum of τ_{diff}. At the same time, the relative detection frame per second is also the minimal. This comes from the fact that the minimum of τ_{diff} creates a large set of difference members and a big coverage region of the region array M^*. As a result, it causes a large size of multiplicands in all convolution layers which takes time to operate. When the τ_{diff} is increased, it is expected that the detection correctness decreases. But the experiment results show that the detection correctness and the relative detection frame per second seem to be constant with the exception when the τ_{diff} is equal to 0.625. At that point, the detection correctness drops abruptly to be only 50% while the relative detection frame per second jumps to about 2.2. This experiment can be concluded that with respect to the video frame the difference threshold τ_{diff} within the range between 0.125 and 0.5625, the detection correctness and the relative detection frame per second seem to be constant: 90% and 1.47, respectively.

5   Conclusions

In this paper, we studied the YOLO object detection which is in the category of fully convolution network. It internally operates by partitioning the input image into non-overlapping square cells. There is a mapping relationship between the input image and the cell within all layers of the network. We drew the relationship which states that **for a particular index of square cells its size can be changed from layer to layer while its index is preserved.** In this way, if a region of change in an input image; the first input feature to the network, can be located and the cells that cover the region can be identified, only the members of the input features within these cells in all layers along the network are required to pay attention; i.e. to be operated by the convolution operation. **The mapping relationship guides us to avoid wasting time to make comparison of the input features; between the frame $t+1$ and t, from layer level to frame level.** When it is employed along with the spatio-temporal property within video frames, it helps accelerating a considerable amount of time. The experiments on the Tiny-YOLO network confirmed that the best relative detection frame per second of 1.47; about 7 fps, was attainable at the detection correctness of 90%. These were benchmarked with the ordinary YOLO object detection on a personal computer without any sophisticate GPUs.

References

1. Redmon, J., Divvala, S., Girshick, R., Farhadi, A.: You only look once: unified, real-time object detection. In: IEEE Conference on Computer Vision and Pattern Recognition (CVPR2016), pp. 779–788. IEEE Press, New York (2016). https://doi.org/10.1109/CVPR.2016.91
2. Redmon, J., Farhadi, A.: YOLO9000: better, faster, stronger. In: IEEE Conference on Computer Vision and Pattern Recognition (CVPR2017), pp. 6517–6525. IEEE Press, New York (2017). https://doi.org/10.1109/CVPR.2017.690
3. Zhang, Q., Zhang, M., Chen, T., Sun, Z., Ma, Y., Yu, B.: Recent advances in convolutional neural network acceleration. Neurocomputing **323**, 37–51 (2019). https://doi.org/10.1016/j.neucom.2018.09.038
4. Bottleson, J., Kim, S., Andrews, J., Bindu, P., Murthy, D.N., Jin, J.: clCaffe: OpenCL accelerated Caffe for convolutional neural networks. In: IEEE International Parallel and Distributed Processing Symposium Workshops (IPDPSW2016), pp 50–57. IEEE Press, New York (2016) https://doi.org/10.1109/IPDPSW.2016.182
5. Cheng, J., Wang, P., Li, G., Hu, Q., Lu, H.: Recent advances in efficient computation of deep convolutional neural networks. Frontiers Inf. Technol. Electron. Eng. **19**, 64–77 (2018). https://doi.org/10.1631/FITEE.1700789
6. Cavigelli, L., Degen, P., Benini, L.: CBinfer: change-based inference for convolutional neural networks on video data. In: Proceedings of the 11th International Conference on Distributed Smart Cameras (ICDSC 2017), pp 1–8. ACM Press, New York (2017). https://doi.org/10.1145/3131885.3131906
7. Xu, M., Zhu, M., Liu, Y., Lin, F.X., Liu, X.: DeepCache: principled cache for mobile deep vision. In: The 24th Annual International Conference on Mobile Computing and Networking (MobiCom), pp 129–144. ACM Press, New York (2018). https://doi.org/10.1145/3241539.3241563

Analyze Facial Expression Recognition Based on Curvelet Transform via Extreme Learning Machine

Sarutte Atsawaruangsuk[✉], Tatpong Katanyukul,
and Pattarawit Polpinit

Computer Engineering Department, Faculty of Engineering,
Khon Kaen University, Khon Kaen, Thailand
sarutte_a@kkumail.com, {tatpong, polpinit}@kku.ac.th

Abstract. This paper aims to investigate the key factors of facial expression recognition based on local curvelet transform for real-time training data. Local curvelet transform (LCT) is the application of curvelet transform that benefits from useful features extracted by curvelet transform and reduces the computation cost of using all curvelet coefficients. The reduction of computation is through calculating the representative features, instead of directly using all curvelet coefficients. The representative features are mean, standard deviation and entropy. This approach has been reported to achieve impressively 0.9445 and 0.9486 accuracy on JAFFE and Cohn-Kanade datasets. However, there are many factors influencing the final performance, in which these factors have not been thoroughly studied. Our investigation has shown that these factors could result up to almost 10% difference and their effects are thoroughly studied.

Keywords: Facial expression recognition · Local Curvelet Transform ·
Extreme Learning Machine

1 Introduction

The facial expression recognition gains interest in various applications, e.g., human-computer interaction, robotic expression, psychological treatment, and virtual reality [1]. The facial expression recognition is task to map from facial image to and appropriate expressing emotion. The expressing emotion is generally categorized into one of the six classes, i.e., happiness, sadness, fear, anger, disgust and surprise [2]. The facial expression recognition has two main parts: feature extraction and learning algorithm.

The facial expressions recognition algorithms can be categorized into feature based and holistic approaches. Feature based approach relies on features, e.g., shape and location of significant facial parts [3], while holistic approach extracts facial features by using a set of filters [4, 5].

The Curvelet Transform (CT) is widely-used method for analysis and synthesis of a digital image in multi-resolution analysis. CT is efficient to capture information in a frequency domain, a spatial domain, and correlation between spatial dimensions. CT is an extension of wavelet transform that has direction analysis. In addition, CT uses the parabola concept to convolute the curvelet to an image in the spatial domain and then

P. Boonyopakorn et al. (Eds.): IC2IT 2019, AISC 936, pp. 148–158, 2020.
https://doi.org/10.1007/978-3-030-19861-9_15

results in a frequency domain. The result of CT is a set of Curvelet Coefficients (CC). Accounting for spatial correlation in polar coordinates, CT can detect curves and lines of a human face better than wavelet transform [6–8].

However, in practice, a size of the CT result—a number of CC size is large, e.g., 28,097 components for a 100×100 image [9]. In order to reduce the dimension of the CT result, the Local Curvelet Transform (LCT) is introduced [8]. The LCT is to apply the CT to local sub-regions of an image. Then, statistical features relating to the local region images are calculated and used as representative features.

The facial features obtained from LCT are then ready for a learning algorithm. An online learning algorithm can facilitate facial expression recognition to reduce the training process. Liang et al. [10] proposed the Online Sequential Extreme Learning Machine (OS-ELM). OS-ELM is implemented based on Extreme Learning Machine (ELM) [11, 12] that allows fast training and good general performance same as ELM, and also provides immediate data updating capability with its online sequential learning.

The combination of LCT and OS-ELM yields the real-time robust data training for facial expression recognition. The LCT provides a lot of features representing an entire facial image. Given the facial features, OS-ELM provides facial expression recognition with accuracy and real-time training capability. LCT and OS-ELM have been shown to have impressive performance [8]. However, they have been used with various settings. Our work is to investigate LCT factors that impact performance and final results. We hypothesize 4 factors: image compression, cropping, statistic features, and normalized features.

This paper is organized as follows: Sect. 2 describes the CT, OS-ELM, and LCT. Section 3 introduces the experimental setup. Section 4 presents the experimental results and discussion, and Sect. 5 provides the conclusion.

2 Related Work

2.1 Digital Curvelet Transform Based on Wrapping Technique

Curvelet transform [6] is defined by the inner product

$$c(j, l, k) := \langle f, \varphi_{j,l,k} \rangle \tag{1}$$

where $\varphi_{j,l,k}$ is the curvelet basis function, f is a signal and j, l, and k are scale, orientation, and position, respectively.

Curvelet Transform (CT) is implemented in the frequency domain. Digital Curvelet Transform (DCT) manages a frequency domain variable ω in the Cartesian coronae based on concentric squares and shears. Figure 1 shows the Cartesian coronae that uses instead of circles in continuous-time CT. The frequency window U_j in Cartesian coronae is defined in the Fourier domain by a radial window W and angular window V. The radial window W is expressed as

$$W_j(\omega) = \sqrt{\Phi_{j+1}^2(\omega) - \Phi_j^2(\omega)}, j \geq 0 \tag{2}$$

where Φ is the product of low-pass one dimensional window [6].

The angular window V is defined as

$$V_j(\omega) = V(2^{j/2}\omega_2/\omega_1) \tag{3}$$

where ω_1 and ω_2 are low-pass one dimensional windows [6]. The Cartesian window $\tilde{U}$ is constructed as

$$\tilde{U}_{j,l}(\omega) = W_j(\omega)V_j(S_\theta\omega) \tag{4}$$

where S_θ is a shear matrix that can defined as $S_\theta = \begin{bmatrix} 1 & \theta \\ \tan\theta & 1 \end{bmatrix}$.

Shear matrix S_θ is used to maintain the symmetry around the original and rotation by $\pm\pi/2$ radiance.

The frequency domain definition of DCT is

$$\varphi_{j,l,k}^D[t_1, t_2] = \tilde{U}_{j,l}[t_1, t_2]e^{-i2\pi[k_1t_1 + k_2t_2]} \tag{5}$$

where $\varphi_{j,k,l}^D$ is a digital curvelet waveform. The discrete curvelet coefficients can be obtained as follows:

$$c^D(j, l, k) = \sum_{0 \le t_1, t_2 < n} f[t_1, t_2]\varphi_{j,l,k}^D[t_1, t_2] \tag{6}$$

where $[t_1, t_2]$ is the index of the image and n is a number of pixels along each dimension.

2.2   ELM

The ELM is the fast training leaning machine proposed by Huang et al. [11]. Let ELM have K hidden nodes, the samples can be written as (x_i, t_i), $i = 1, 2, \ldots, N$ where $x_i \in \mathbb{R}^M$ and $t_i \in \mathbb{R}^C$, M and C are numbers of input and output dimensions, respectively. Let $\mathbf{X} = \{x_1, x_2, \ldots, x_N\}$ be the input samples and $\mathbf{T} = \{t_1, t_2, \ldots, t_N\}$ be the output samples. ELM can be written into the least square form as follows:

$$\beta = \mathbf{H}^\dagger \mathbf{T} \tag{7}$$

where $\beta = [\beta_{j1}, \beta_{j2}, \ldots, \beta_{jC}]^T$, $j = 1, 2, \ldots, K$, is a matrix of output weights. $\mathbf{H}^\dagger$ is the inverse of $\mathbf{H}$ from the Moore-Penrose pseudo inverse:

$$\mathbf{H} = [h_{ij}] = \sum_{i=1}^{N} w_j x_i + b_j, \tag{8}$$

where input weights w_j and biases b_j be randomly generated.

2.3 OS-ELM

Liang et al. [10] proposed Online Sequential extreme Learning Machine (OS-ELM) that can be trained in an online manner. Calculation of the OS-ELM can be summarized as follows:

Given the previously trained data $(\mathbf{X}_{k-1}, \mathbf{T}_{k-1})$, the newly arrived data $(\mathbf{X}_k, \mathbf{T}_k)$, and the number of hidden nodes K is less than or equal to the number of samples N, the output weights are defined as

$$\beta_k = \mathbf{K}_k^{-1} \begin{bmatrix} \mathbf{H}_{k-1} \\ \mathbf{H}_k \end{bmatrix}^T \begin{bmatrix} \mathbf{T}_{k-1} \\ \mathbf{T}_k \end{bmatrix} \tag{9}$$

where and k is an index of newly arrived data and $k-1$ is index of previously trained data.

$$\mathbf{K}_k = \begin{bmatrix} \mathbf{H}_{k-1} \\ \mathbf{H}_k \end{bmatrix}^T \begin{bmatrix} \mathbf{H}_{k-1} \\ \mathbf{H}_k \end{bmatrix} \tag{10}$$

Equation 10 can be rewritten into the terms of $\mathbf{K}_{k-1}$ for sequential learning as

$$\mathbf{K}_k = \mathbf{K}_{k-1} + \left(\mathbf{H}_k^T \mathbf{H}_k \right). \tag{11}$$

Thus, with Eqs. 9 and 11, we have

$$\beta_k = \beta_{k-1} + \mathbf{K}_k^{-1} \mathbf{H}_k^T \left(\mathbf{T}_k - \mathbf{H}_k \beta_{k-1} \right). \tag{12}$$

In order to calculate $\mathbf{K}_k^{-1}$, Eq. 11 can be reformulated based on the Woodbury formula [13] to be

$$\mathbf{P}_k = \mathbf{P}_{k-1} + \mathbf{P}_{k-1} \mathbf{H}_k^T \left(I + \mathbf{H}_k^T \mathbf{P}_{k-1} \mathbf{H}_k \right)^{-1} \mathbf{H}_k \mathbf{P}_{k-1}, \tag{13}$$

where $\mathbf{P}_k = \mathbf{K}_k^{-1}$.

2.4 Facial Expression Recognition Based on Local Curvelet Transform

Local Curvelet Transform (LCT) [8] is designed to reduce a number of Curvelet Coefficients (CC). LCT can be used to provide a concise set of features to a learning model, which is configured to classify facial expression. Facial expression recognition based on LCT can be summarized as follows:

(1) Detecting face. Viola-Jones face detection [14] can be used for face detection.
(2) Cropping the image. Images are cropped and resized to 128 × 128.
(3) Applying histogram equalization [15].

(4) Separating the image into sub-regions. The image is separated into 64 sub-regions each having size of 16×16.
(5) Applying curvelet transform to each sub-region.
(6) Calculating statistical features. Curvelet coefficients of the first scale, i.e., in Eq. 6, are used to calculate mean, entropy and standard deviation of each region.
(7) Using the statistical features to be the input of a learning machine.

3 Experimental Setup

The JApanese Female Facial Expression (JAFFE) [16] and Cohn-Kanade [17] datasets are used to evaluate the performance of the LCT with ELM and with OS-ELM. All methods are implemented using MATLAB version R2014a on a computer with Core i3 3.40 GHz Ram 4.00 GB. Our code can be downloaded from https://github.com/sarutte/ LCT/ELM. Our experiment uses LCT implementation from http://www.curvelet.org/ software.html. Our experiment uses OS-ELM implementation from http://www.ntu. edu.sg/home/egbhuang/elm_codes.html. The leave-one-subject-out method is used to validate the performance of ELM and OS-ELM. The numbers of the hidden nodes of ELM and OS-ELM are varied in the range of [1,500]. The details of each dataset are described as follows:

JAFFE dataset [16] contains 213 facial expression images. The facial expression images are posted by 10 Japanese females. Each subject performs 7 expressions and each expression has two to four different images with the resolution 256×256 pixels.

The extended Cohn-Kanade dataset (CK+) [17] contains 10,588 facial expression images. Our experiment selects only images that have the associated emotion labels. That results in 1,100 facial expression images from 89 individuals.

To investigate LCT related setting, we examine the following factors:

1. Factor 1 (Image compression). No compression (Original image, Factor 1a) and JPG compression (Factor 1b) are compared. JPG compression reduces some visual information that may be a noise in an image. We speculate that it may help the recognition performance.
2. Factor 2 (Cropping). An image without cropping (Factor 2a) and the cropping of the unnecessary part (Factor 2b) are evaluated. We speculate that cropping the unnecessary part may increase the accuracy.
3. Factor 3 (Statistic features). We evaluate two approaches for preparing input features: (1) Local Curvelet Coefficients (LCC) and (Factor 3a) (2) statistics of the coefficients (Factor 3b). Statistical features may provide overall characteristics of the coefficients.
4. Factor 4 (Normalization range). Normalization ranges [0,1] (Factor 4a) and [−1,1] (Factor 4b) are compared. A wider range may be more suitable for a large number of feature dimensions.

4 Experimental Results and Discussion

4.1 Facial Expression Recognition Based on Local Curvelet Transform

In order to provide a base case, we repeat the facial expression recognition based on local curvelet transform [8]. Although we strictly followed Uçar et al. [8], we obtain quite different results as shown in Table 1.

Table 1. Accuracy of LCT in our experiment and LCT paper [8] on JAFFE and Cohn-Kanade datasets

Dataset	JAFFE		Cohn-Kanade	
Machine	ELM	OS-ELM	ELM	OS-ELM
Uçar et al. [7]	0.9545	0.9445	0.9464	0.9486
Our results	0.7009	0.7499	0.8653	0.8658
Difference	0.2536	0.1946	0.0811	0.0828

Table 2. Accuracy of the JAFFE dataset

Machine	Normalized range	Original image type				JPG image			
		Feature		LCC		Feature		LCC	
		No crop	Crop	No crop	Crop	No crop	Crop	No crop	Crop
ELM	[0,1]	0.7009	0.7240	0.7093	0.7101	0.7002	0.7331	0.7140	0.7044
	[−1,1]	0.6802	**0.7567**	0.7016	0.7376	0.7197	0.7203	0.7234	0.7476
OS-ELM	[0,1]	0.7499	0.7522	0.7374	0.7579	0.7329	0.7427	0.7374	0.7431
	[−1,1]	0.7335	0.7504	0.7440	**0.7577**	0.7375	0.7559	0.7491	**0.7577**

Table 3. Accuracy of the Cohn-Kanade dataset

Machine	Normalized range	Original image type				JPG image			
		Feature		LCC		Feature		LCC	
		No crop	Crop	No crop	Crop	No crop	Crop	No crop	Crop
ELM	[0,1]	0.8653	0.8715	0.8212	0.8678	0.8580	0.8705	0.8218	0.8657
	[−1,1]	0.8461	0.8660	0.8163	0.8690	0.8386	0.8642	0.8208	**0.8726**
OS-ELM	[0,1]	0.8658	0.8744	0.8008	0.8583	0.8568	**0.8788**	0.8061	0.8517
	[−1,1]	0.8454	0.8631	0.8159	0.8696	0.8454	0.8565	0.8367	0.8639

Table 4. Average accuracies on JAFFE and Cohn-Kanade datasets

Machine	Normalized range	Original image type				JPG image			
		Statistical feature		All LCC		Statistical feature		All LCC	
		No crop	Crop	No crop	Crop	No crop	Crop	No crop	Crop
ELM	[0,1]	0.7831	0.7977	0.7652	0.7890	0.7791	0.8018	0.7679	0.7851
	[−1,1]	0.7631	**0.8113**	0.7590	0.8033	0.7792	0.7923	0.7721	0.8101
OS-ELM	[0,1]	0.8079	0.8133	0.7691	0.8081	0.7949	0.8108	0.7718	0.7974
	[−1,1]	0.7895	0.8067	0.7800	**0.8137**	0.7915	0.8062	0.7929	0.8108

In Tables 2, 3 and 4, the bold letters indicate the best accuracy for each machine. Tables 2 and 3 can be summarized into Table 4. Table 4 shows the average accuracy of ELM and OS-ELM on JAFFE and Cohn-Kanade datasets. The results show that ELM has the best accuracy (0.8113) when working with an original image type (Factor 1a), cropping (Factor 2b), the statistical features (Factor 3b), and normalization range [−1,1] (Factor 4b). As for OS-ELM results, OS-ELM has the best accuracy (0.8137) when working with an original image type (Factor 1a), cropping (Factor 2b), LCC (Factor 3b) and normalization range [−1,1] (Factor 4b). Although the best performing setting of OS-ELM (at 0.8137) uses all LCCs, the second best with comparable accuracy (at 0.8133) uses only statistical features. Given a comparable accuracy, using statistical features is more favorable for its cheaper computation cost.

Table 5. Average different accuracies of ELM and OS-ELM on JAFFE and Cohn-Kanade datasets with different factors

Factors	ELM	OS-ELM
Image compression type: original type VS JPG type		
- No crop, stat. features, [0,1]	0.0040 ± 0.0223	0.0130 ± 0.0349
- No crop, all LCC, [0,1]	**−0.0027 ± 0.0600**	**−0.0026 ± 0.0402**
- No crop, stat. features, [−1,1]	**−0.0161 ± 0.0323**	**−0.0020 ± 0.0309**
- No crop, all LCC, [−1,1]	**−0.0131 ± 0.0352**	**−0.0129 ± 0.0358**
- Crop, stat. features, [0,1]	**−0.0041 ± 0.0299**	0.0025 ± 0.0303
- Crop, all LCC, [0,1]	0.0039 ± 0.0351	0.0107 ± 0.0266
- Crop, stat. features, [−1,1]	0.0190 ± 0.0481	0.0006 ± 0.0291
- Crop, all LCC, [−1,1]	**−0.0068 ± 0.0253**	0.0029 ± 0.0291

(continued)

Table 5. (*continued*)

Cropping: no cropping VS cropping		
- Original type, stat. features, [0,1]	**−0.0147 ± 0.0481**	**−0.0054 ± 0.0221**
- Original type, all LCC, [0,1]	**−0.0238 ± 0.0341**	**−0.0389 ± 0.0568**
- Original type, stat. features, [−1,1]	**−0.0482 ± 0.0214**	**−0.0173 ± 0.0410**
- Original type, all LCC, [−1,1]	**−0.0443 ± 0.0378**	**−0.0337 ± 0.0551**
- JPG type, stat. features, [0,1]	**−0.0227 ± 0.0268**	**−0.0159 ± 0.0384**
- JPG type, all LCC, [0,1]	**−0.0172 ± 0.0480**	**−0.0257 ± 0.0372**
- JPG type, stat. features, [−1,1]	**−0.0131 ± 0.0590**	**−0.0147 ± 0.0326**
- JPG type, all LCC, [−1,1]	**−0.0380 ± 0.0362**	**−0.0179 ± 0.0414**
Feature: statistic feature VS All LCC feature		
- Original type, no crop, [0,1]	0.0178 ± 0.0418	0.0387 ± 0.0387
- Original type, crop, [0,1]	0.0087 ± 0.0367	0.0052 ± 0.0297
- Original type, no crop, [−1,1]	0.0041 ± 0.0346	0.0095 ± 0.0432
- Original type, crop, [−1,1]	0.0080 ± 0.0220	**−0.0070 ± 0.0488**
- JPG type, no crop, [0,1]	0.0112 ± 0.0496	0.0231 ± 0.0333
- JPG type, crop, [0,1]	0.0167 ± 0.0305	0.0133 ± 0.0384
- JPG type, no crop, [−1,1]	0.0071 ± 0.0284	**−0.0014 ± 0.0384**
- JPG type, crop, [−1,1]	**−0.0178 ± 0.0651**	**−0.0046 ± 0.0309**
Normalization range: [0,1] VS [−1,1]		
- Original type, no crop, stat. features	0.0199 ± 0.0290	0.0184 ± 0.0248
- Original type, crop, stat. features	**−0.0136 ± 0.0258**	0.0066 ± 0.0294
- Original type, no crop, all LCC	0.0062 ± 0.0411	**−0.0108 ± 0.0327**
- Original type, crop, all LCC	**−0.0143 ± 0.0245**	**−0.0056 ± 0.0329**
- JPG type, no crop, stat. features	**−1E-04 ± 0.0280**	0.0034 ± 0.0214
- JPG type, crop, stat. features	0.0095 ± 0.0509	0.0046 ± 0.0272
- JPG type, no crop, all LCC	**−0.0042 ± 0.0443**	**−0.0211 ± 0.0471**
- JPG type, crop, all LCC	**−0.0250 ± 0.0313**	**−0.0134 ± 0.0201**

Table 5 shows average different accuracies of ELM and OS-ELM on JAFFE and Cohn-Kanade datasets with different factors. The numbers indicate difference between accuracy of the first option and an accuracy of the second option. For example, the cropping option has been shown to be a preferable option for all other settings. An image with cropping leads to a higher accuracy than an image without cropping, given the same other factors.

4.2   Generality of LCT Based Facial Expression Recognition

Some previous conception is that LCT based facial expression recognition is general in the sense that given a facial image, a system can recognize an emotion accurately, regardless of whether the given image belongs to an individual whose face has been in the training process [8].

To test this conception, we compare performance measure using Leave-1-out method where no facial example of the test individuals appears in the training process to measure using the methods, as follows:

1. Leave-one-subject-out method. The testing data are composed of images of one subject out of the total 10 subjects. The remaining images are used for the training data. For JAFFE, one subject associates to one individual. For Cohn-Kanade, one subject associates to 8–9 individuals.
2. Leave-1-subject-out+7 method. This method is similar to the leave-one-subject-out method, but it adds each image for every emotion (all 7 emotions) of the test individual.
3. Train-1-emotion-image method. It uses each image for each emotion of all individuals for training and the rest for test.
4. Test-1-emotion-image method. It uses each image for each emotion of all individuals for test and the rest for training.
5. K-fold cross-validation method.

Table 6 shows the accuracies of ELM and OS-ELM on the JAFFE dataset and the Cohn-Kanade dataset with various evaluation methods. All other evaluation methods results in considerably higher accuracies than the leave-1-out method. This apparently implies that if the recognition accuracy is higher when using more image of the test subject in the training process. This contradicts the common notion that LCT is general in the sense that LCT can bring out striking emotion features, regardless of specific individual faces.

In order to quantify a relation between a number of emotions of the test subject in the training process and the final accuracy (Fig. 1), we experimented a few extra methods. The extra methods can be commonly defined as Leave-1-subject-out+n. The leave-1-subject-out+n is similar to the leave-one-subject-out method, but it adds each image for each of selected n emotions of the test individual. Therefore the leave-1-subject-out is equivalent to leave-1-subject-out+0.

Regarding the issue of generalization of LCT, the causing factors might be (1) each person may have posed a different look for the same emotion. (2) The facial images under study are not aligned or angled, e.g., slightly right-turn face, slightly down-tilt face. This kind of off-angle images is prevalent especially in JAFFE database. (3) The facial shape may also affect the classification.

LCT calculates Local Curvelet Coefficient (LCC) from lines and curves of a face image directly. LCT is effective for detecting the low-level features. Therefore, it may be beneficial for having an additional mechanism to emphasize the relevant or filter out the irrelevant low-level features obtained from LCT.

Table 6. Accuracies of JAFFE and Cohn-Kanade datasets with different validation methods

Machine	Dataset	Leave-1-subject-out	Leave-1-out +7	Train-1-emotion-image method	Test-1-emotion-image method	K-fold
ELM	JAFFE	0.7009	0.9871	0.9220	0.9565	1.0000
OS-ELM	JAFFE	0.7499	0.9938	0.8298	0.9275	1.0000
ELM	CK+	0.8653	0.9764	0.9770	0.9818	0.9882
OS-ELM	CK+	0.8658	0.9848	0.9721	0.9782	0.9900

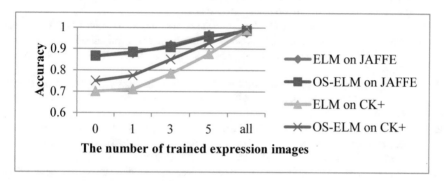

Fig. 1. The accuracies of ELM and OS-ELM on JAFFE and Cohn-Kanade datasets while add trained expression image.

5 Conclusion

Our experiment investigates the factors related to applying LCT to facial expression recognition. Our findings are: (1) ELM and OS-ELM are suitable to use with a statistical summary of LCT features. (2) Cropping image contributes to higher accuracy. It removes unnecessary parts from an image. (3) Training with more images of test subject improves accuracy. This implies that LCT does not have the generalization.

References

1. Anderson, K., Mcowan, P.: A real-time automated system for the recognition of human facial expressions. IEEE Trans. Syst. Man Cybern. Part B (Cybern.) **36**, 96–105 (2006). https://doi.org/10.1109/TSMCB.2005.854502
2. Ekman, P., Friesen, W.V.: Constants across cultures in the face and emotion. J. Pers. Soc. Psychol. **17**, 124–129 (1971). https://doi.org/10.1037/h0030377
3. Tian, Y.-I., Kanade, T., Cohn, J.: Recognizing action units for facial expression analysis. IEEE Trans. Pattern Anal. Mach. Intell. **23**, 97–115 (2001). https://doi.org/10.1109/34.908962

4. Wen, Z, Huang, T.S.: Capturing subtle facial motions in 3D face tracking. In: 9th IEEE International Conference on Computer Vision. IEEE Press, France (2003). https://doi.org/10.1109/iccv.2003.1238646
5. Gu, W., Xiang, C., Venkatesh, Y., Huang, D., Lin, H.: Facial expression recognition using radial encoding of local Gabor features and classifier synthesis. Pattern Recognit. **45**, 80–91 (2012). https://doi.org/10.1016/j.patcog.2011.05.006
6. Candès, E.J., Donoho, D.L.: New tight frames of curvelets and optimal representations of objects with piecewise C2 singularities. Commun. Pure Appl. Math. **57**, 219–266 (2003). https://doi.org/10.1002/cpa.10116
7. Mohammed, A., Minhas, R., Wu, Q.J., Sid-Ahmed, M.: Human face recognition based on multidimensional PCA and extreme learning machine. Pattern Recognit. **44**, 2588–2597 (2011). https://doi.org/10.1016/j.patcog.2011.03.013
8. Uçar, A., Demir, Y., Güzeliş, C.: A new facial expression recognition based on curvelet transform and online sequential extreme learning machine initialized with spherical clustering. Neural Comput. Appl. **27**, 131–142 (2014). https://doi.org/10.1007/s00521-014-1569-14
9. Tang, M., Chen, F.: Facial expression recognition and its application based on curvelet transform and PSO-SVM. Optik **124**, 5401–5406 (2013). https://doi.org/10.1007/978-3-030-01174-1_45
10. Liang, N.-Y., Huang, G.-B., Saratchandran, P., Sundararajan, N.: A fast and accurate online sequential learning algorithm for feedforward networks. IEEE Trans. Neural Netw. **17**, 1411–1423 (2006). https://doi.org/10.1109/TNN.2006.880583
11. Huang, G.-B., Zhu, Q.-Y., Siew, C.-K.: Extreme learning machine: theory and applications. Neurocomputing **70**, 489–501 (2006). https://doi.org/10.1016/j.neucom.2005.12.126
12. Huang, G.-B., Zhou, H., Ding, X., Zhang, R.: Extreme learning machine for regression and multiclass classification. IEEE Trans. Syst. Man Cybern. Part B (Cybern.) **42**, 513–529 (2012). https://doi.org/10.1109/TSMCB.2011.2168604
13. Golub, G.H., Van Loan, C.F.: Matrix Computations. The Johns Hopkins University Press, Baltimore (2013)
14. Viola, P., Jones, M.: Robust real-time face detection. In: 8th IEEE International Conference on Computer Vision, ICCV 2001. IEEE Press, Canada (2001). https://doi.org/10.1109/iccv.2001.937709
15. Acharya, T., Ray, A.K.: Image Processing: Principles and Applications. Wiley, Hoboken (2005)
16. Facial Expression Database: Japanese Female Facial Expression (JAFFE) Database. http://www.kasrl.org/jaffe.html
17. Lucey, P., Cohn, J.F., Kanade, T., Saragih, J., Ambadar, Z., Matthews, I.: The extended Cohn-Kanade Dataset (CK): a complete dataset for action unit and emotion-specified expression. In: 2010 IEEE Computer Society Conference on Computer Vision and Pattern Recognition - Workshops. IEEE Press, California (2010). https://doi.org/10.1109/cvprw.2010.5543262

Ensemble Model for Segmentation of Lateral Ventricles from 3D Magnetic Resonance Imaging

Akadej Udomchaiporn$^{(\boxtimes)}$, Khitichai Lertrungwichean,
Pokpakorn Klinkasen, and Chawanwut Nuchprasert

Department of Computer Science, Faculty of Science,
King Mongkut's Institute of Technology Ladkrabang, Bangkok 10520, Thailand
{akadej.ud, 57050186, 57050214, 57050205}@kmitl.ac.th

Abstract. The paper proposes an ensemble model to segment lateral ventricles from 3D Magnetic Resonance Imaging (MRI) brain scan. Thresholding and Active Contour techniques combining with a noise removal method were applied to segment lateral ventricles from the brain images. The experiments were conducted by segmenting 73 MRI brain scans using our proposed model and then comparing their volumes to those using manual model conducted by an expert. The experimental results indicated that the proposed model segmented lateral ventricles as excellent as the manual model in terms of accuracy but outperformed the manual model in terms of time performance. The contribution of the paper is that the segmented lateral ventricles can be used for further analysis such as medical condition classification.

Keywords: 3D Magnetic Resonance Imaging · MRI brain scan ·
Lateral ventricles · Image segmentation · Image processing

1 Introduction

Magnetic Resonance Imaging (MRI) came into prominence in the last 50 years. It is a medical imaging technique widely used to investigate the anatomy and physiology of the human (or animal) body. It allows the user to see inside without the need for dissection [1]. MRI uses magnetic fields and radio waves to generate high quality and detailed computerized images of the inside of the body. Recently, MRI has advanced beyond a 2D imaging technique to a 3D imaging technique which can produce a very detailed image of the scanned object that can be viewed as a 3D image. Nowadays, MRI is commonly used to examine the joints, abdomen, pelvis, spine, and brain of the subjects. An MRI scanner is a cylinder shape that is open at both ends. The patient lies horizontally on a bed that can be moved into the scanner cylinder. The patient needs to remain still during the scan in order to produce the best quality MR image and the process takes approximately 20 min. An example of an MRI brain scan is shown in Fig. 1.

The focus of this paper is directed at 3D MRI brain scan, specifically at lateral ventricles. The ventricles are fluid-filled open spaces at the centre of the brain [2]. There are, in fact, four ventricles in a human brain, but only lateral ventricles are considered in

© Springer Nature Switzerland AG 2020
P. Boonyopakorn et al. (Eds.): IC2IT 2019, AISC 936, pp. 159–168, 2020.
https://doi.org/10.1007/978-3-030-19861-9_16

the paper. The reason is that the lateral ventricles can be relatively clearly distinguished within a 3D MRI brain scan. An example of a 3D MRI brain scan is given as Fig. 2 where the "lateral ventricles" are the dark areas at the centre of the brain.

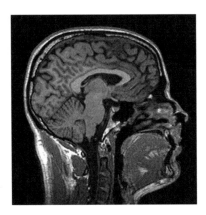

Fig. 1. Example of an MRI brain scan

The size and shape of organs in human brains have been observed to be correlated to some medical conditions or diseases such as Schizophrenia, Multiple Sclerosis and Epilepsy. It is also related to various laterised behavior in people such as handedness. Several studies indicate that the size of the ventricular system (including lateral ventricles) in humans is correlated to some medical conditions. For example, Thom [3] was the first to observe ventricular enlargement in Epilepsy patients. He reported lateral ventricle dilation following postmortem examinations of patients with idiopathic Epilepsy. The ventricular system has also been associated with other disorders such as Schizophrenia, Multiple Sclerosis, Alzheimers and Parkinsons [4]. Significant ventricular enlargement has been associated with Alzheimers [5]. Similarly, an explorative study of Parkinsons suggests that ventricular enlargement has been associated with early cognitive impairment. Some of the work has also shown that ventricular enlargement of the lateral ventricles including the third ventricle was associated with neuropsychological functions in advanced non-demented Pakinson patients [6]. Moreover, the presence of ventricular enlargement in both Epilepsy and Schizophrenia has indicated a common neurodevelopmental mechanism that predisposes to Epileptogenesis and Schizophrenia [7]. In a study of severe myoclonic Epilepsy in infancy, 6 out of 13 patients investigated exhibited moderate ventricular enlargement [8]. Age accelerated changes in Epilepsy participants (in comparison with healthy people) have been seen in the lateral ventricles, whereas largely comparable patterns of age-related changes were seen across other regions of interest in the brain [9].

The paper therefore proposed the model to segment lateral ventricles from a 3D MRI brain scan. The contribution of the paper is that the segmented lateral ventricles can be used for further analysis such as medical condition classification. The structure of the paper is organized as follows. Section 2 introduces some related work. In Sect. 3, the ensemble segmentation model is described. The experimental results are shown in Sect. 4. Evaluation including discussion are presented in Sect. 5. Finally, the paper is summarised in Sect. 6 with conclusion of the main findings.

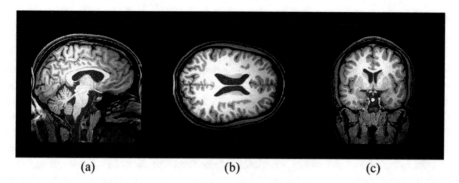

Fig. 2. Example of a 3D MRI brain scan: (a) Sagittal plane, (b) Transverse plane, and (c) Coronal plane

2   Related Work

Some of the published work on the segmentation of brain images has common aspects with the work described in this paper. For example, Elsayed et al. [10, 11] proposed a Modified Spectral Segmentation algorithm, founded on a multiscale graph decomposition to segment the corpus callosum (another organ in MRI brain scans). Experimentation was conducted using a dataset of 76 MRI brain scans; the results indicated that their algorithm was able to detect the corpus callosum more accurate than those when using existing segmentation techniques. However, the works were directed at 2D data (specifically, the mid-sagittal slice of an MRI volume). Next, Long [12] presented a technique to segment lateral ventricles including the third ventricle. This segmentation technique automatically adjusted the most suitable threshold value for each brain image. It was argued that this tended to produce more accurate result. However, the work described in [12] had time consuming in dynamic thresholding process and had no evaluation concerning the segmentation. Some other reported work on MRI brain scan segmentation can be found in [13] where M. Rousson et al. proposed a manual approach to extract brain ventricles from a 3D MRI brain scan. The approach was implemented using active shape models [14] and level set methods [15]. However, like in the case of [12], no evaluation of the segmentation was included. In 2017, Kulworatit et al. proposed a semi-automatic model to segment lateral ventricles from a 3D MRI brain scan [16]. Thresholding technique and Flood Fill algorithm were applied to identify shape and size of the lateral ventricles. The experimental results indicated that the model was able to segment lateral ventricles accurately comparing to those manually segmented by an expert. However, the model had limitations in some of the cases that the boundary of the lateral ventricles was not clear. Hence, some of manual work had to be done at a stage between thresholding and flood-fill process. Finally, Akadej et al. [17] presented a procedure to segment lateral ventricles from a 3D MRI scan using Bounding Box segmentation technique. Although the results were promising comparing to those manually segmented by an expert, the limitation was that the method had some problems with the case of extremely small lateral ventricles.

3 The Proposed Model

To identify the lateral ventricles from a 3D MRI brain scan, the ensemble model is proposed using some of the existing models such as Thresholding [18] and Active contour model [19]. The input of the process is a 3D MRI brain scan, consisting of 256 image slices (256 × 128 pixels per slice). There are three planes: Sagittal, Coronal and Transverse, in a 3D MRI brain scan. The middle slices for each plane are shown in Fig. 2. However, only image slices in Transverse plane were selected to be processed with respect to this research. The challenging issues of the segmentation process are that the boundary of the lateral ventricles for some image slices is unclear. In addition to that, some part of the third ventricle overlaps with the boundary of the lateral ventricles for some image slices. By applying the proposed model, the issues are solved. The segmentation process can be separated into three sub-processes:

1. Image Enhancement
2. Active Contour
3. Third Ventricle Removal

Each sub-process is explained in detail as follows.

3.1 Image Enhancement

This step is to adjust contrast of the image slices using Contrast-Limited Adaptive Histogram Equalization (CLAHE) [20] in order to improve the contrast of the image. Hence, the boundary of the brain organs in the images is clearer. The example of this step is illustrated in Fig. 3. Next, Thresholding technique [18] is used to convert grey scale images into black and white images. Note that the thresholding value used in the experiment is 0.25. The image is then complemented (swap black and white pixels). Finally, the border of the image (skull) is removed in order to focus on the region of interest. The example of these steps is shown in Fig. 4.

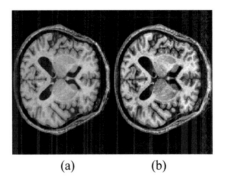

(a)   (b)

Fig. 3. Example of a brain image (a) before and (b) after using CLAHE technique

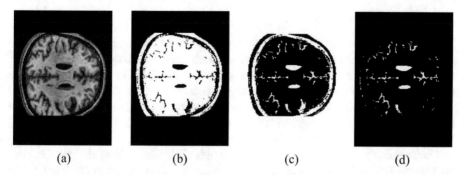

Fig. 4. Example of a brain image in image enhancement process (a) original image, (b) image thresholding, (c) image complement, and (d) skull border removal

3.2   Active Contour

The objective of this step is to identify the boundary of the lateral ventricles using Active Contour model (or Snakes) [19]. The technique is applied for delineating the lateral ventricle outline from any other organs (noises). The example of these steps is shown in Fig. 5.

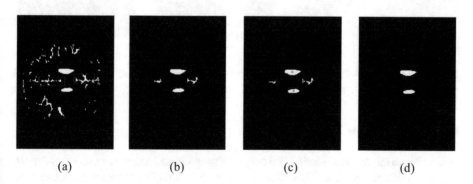

Fig. 5. Example of a brain image in Active Contour process (a) original image, (b) with noises, (c) lateral ventricle identification, and (d) removing noises and starting active contour

According to Fig. 5, the process starts by removing point clouds that their area is less than 3 voxels (Fig. 5(b)). After this step, there are still some noises in the image. Hence, a noise removal technique was applied by identifying lateral ventricles (top and bottom point clouds shown in Fig. 5(c)) and then removing any other remaining point clouds in the image (Fig. 5(d)). Finally, the Active Contour algorithm starts crawling to the next image slices in order to identify the boundary of the lateral ventricles. The process repeats until all of the image slices are crawled and then the volume of the lateral ventricles can be generated from all of the image slices in the plane.

3.3   Third Ventricle Removal

This step is to remove the third ventricle (considered as noises) which might appear in some image slices in some cases. The third ventricle appearing in the lateral ventricles is shown in Fig. 6(a). In this step, the manual process is needed to explore whether the third ventricle appears in the segmented lateral ventricle images. If it happens, the third ventricle removal process is required. The process starts by manually identifying the first image slice that the third ventricle appears and then the algorithm identifies the lateral (left and right) ventricles by considering only top and bottom point clouds (Fig. 6(b)) and consequently remove any other point clouds (Fig. 6(c)). Then, the process goes back to the Active Contour model described in the previous section in order to remove all parts of the third ventricle appearing in the consecutive image slices.

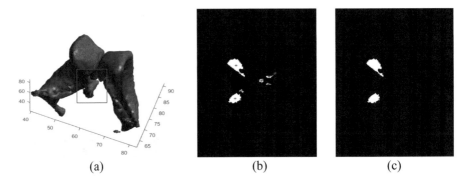

(a)   (b)   (c)

Fig. 6. Example of a brain image (a) with third ventricle (considered as noises), (b) lateral ventricle identification, and (c) removing third ventricle

4   Experimental Results

According to the previous section, the output of the proposed model is a point cloud of the segmented lateral ventricles. The point cloud is a set of coordinate points $(x, y$ and $z)$ in the three-dimensional Cartesian coordinate system. The point cloud can be transformed into a 3D image by using various existing software products. An example of left and right ventricle 3D image generated using Matlab [21] is shown in Fig. 7.

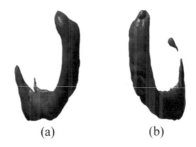

(a)   (b)

Fig. 7. Example of extracted (a) left and (b) right ventricles generated using Matlab

To evaluate the proposed 3D segmentation model, experiments were conducted using a 3D MRI dataset consisting 73 MRI 3D brain scan of the healthy subjects. The dataset used in the experiments is supported by Magnetic Resonance and Image Analysis Research Centre (MARIARC) at University of Liverpool. All of the 73 MRI brain scans in the dataset had been manually segmented and calculated by a domain expert[1]. The process of manual segmentation of the ventricle volumes was conducted by manually marking pixels which are in the area of lateral ventricles on a computer screen, slice by slice, in Coronal plane. The number of selected pixels on each slice was then summed and considered to represent the 3D volume of the given ventricle in terms of voxels. The results of the comparison of the volumes are presented in Figs. 8 and 9, where the volumes obtained using the manual method (blue line) and the proposed method (red line) are plotted. Figure 8 shows the result of left ventricle and Fig. 9

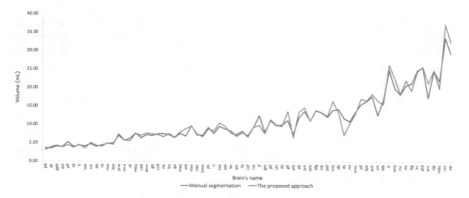

Fig. 8. Comparisons between left ventricle volumes obtained using manual estimation and proposed model

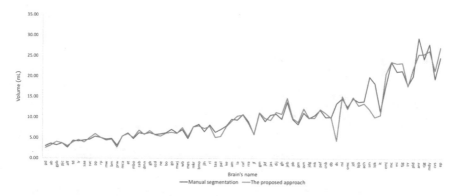

Fig. 9. Comparisons between right ventricle volumes obtained using manual estimation and proposed model

---

[1] Dr. Vanessa Sluming, a leading neuroimaging scientist at University of Liverpool.

shows the result of right ventricle. For each plot, the brain's names, sorted according to the associated ventricle volume size, is listed in the *x*-axis, while the volume sizes (in milliliter: mL) are shown in the *y*-axis.

From Figs. 8 and 9, it can be seen that all the segmented volume values obtained from the proposed model were marginally different to the manually calculated volume values. It is arguable that the manually calculated volumes do not provide a real standard may be flawed due to human error. However, they did provide a benchmark. Closer investigation of the figures is described in the next section.

5   Discussion

To give an insight into the results obtained, the differences in volume size (mL) between the manual and the proposed algorithm are presented using Bland-Altman plots in Figs. 10 and 11. A Bland-Altman plot is a statistical data plotting technique for assessing the "agreement" or "not agreement" between two methods of measurement [22]. The mean difference (bias estimation) and the range of agreement, calculated as the mean difference between −2SD and +2SD, are represented by the continuous horizontal lines.

From Figs. 10 and 11, it can be seen that 92% (67 out of 73) of the left ventricle volume difference and 96% (70 out of 73) of the right ventricle volume difference between the manually calculated volume and the volume collected using the proposed model are in the acceptable boundary (red lines). The interpretation is that the manually segmented method conducted by the expert and the proposed segmentation method produced indifferently promising results. However, the manually segmented method by the expert is time consuming and is outperformed by the proposed approach in terms of time performance.

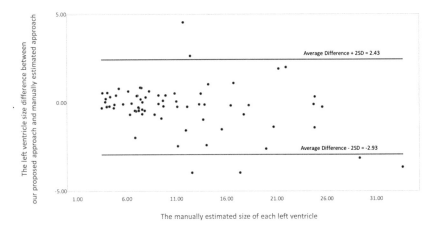

Fig. 10. Bland-Altman plot comparing left ventricle volumes obtained using manual calculation and the proposed model

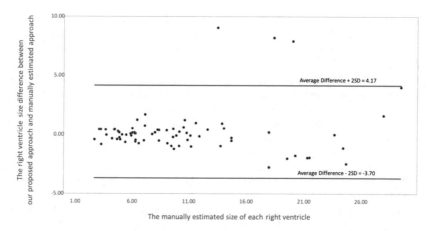

Fig. 11. Bland-Altman plot comparing right ventricle volumes obtained using manual calculation and the proposed model

6 Conclusion

The paper proposed a segmentation model to identify the volume of lateral ventricles from 3D MRI brain scan. Some existing methods such as Thresholding, Active Contour and noise removal were applied accompanied with some more techniques, thus "ensemble" model. The experimental results indicated that the proposed ensemble model is as excellent as the manual model in terms of effectiveness (accuracy) but outperforms the manual model in terms of efficiency (time performance). The only disadvantage of the proposed model is that it requires some manual work to select a "seed" or starting point for Active Contour model (in Sect. 3.2) and to check the third ventricle (considered as noises) including to identify the seed to start the Active Contour process again in order to remove the noises (in Sect. 3.3). However, when comparing to the previous related work, the proposed model outperforms most of the work in terms of efficiency. The fully-automatic process will be conducted and experimented as our future work together with the analysis of some medical conditions using the segmented lateral ventricles.

References

1. Edelman, R., Warach, S.: Magnetic resonance imaging. N. Engl. J. Med. **328**(10), 708–716 (1993)
2. Lin, J.L., Mula, M., Hermann, B.P.: Uncovering the neurobehavioural comorbidities of epilepsy over the lifespan. Lancet **380**(9848), 1180–1192 (2012)
3. Thom, D.: Dilatation of the lateral ventricles as a common brain lesion in epilepsy. J. Nerv. Ment. Dis. **46**(5), 355–358 (1917)
4. Jackson, D.C., Irwin, W., Dabbs, K., Lin, J.J., Jones, J.E., Hsu, D.A., Stafstrom, C.E., Seidenberg, M., Hermann, B.P.: Ventricular enlargement in new-onset pediatric epilepsies. Epilepsia **52**(12), 2225–2232 (2011)

5. Apostolova, L.G., Green, A.E., Babakchanian, S., Hwang, K.S., Chou, Y., Toga, A.W., Thompson, P.M.: Hippocampal atrophy and ventricular enlargement in normal aging, mild cognitive impairment and Alzheimer's disease. Alzheimer Dis. Assoc. Disord. **26**(1), 17 (2012)
6. Dalaker, T.O., Zivadinov, R., Ramasamy, D.P., Beyer, M.K., Alves, G., Bronnick, K.S., Tysnes, O.B., Aarsland, D., Larsen, J.P.: Ventricular enlargement and mild cognitive impairment in early Parkinson's disease. Mov. Disord. **26**(2), 297–301 (2011)
7. Cascella, N.G., Schretlen, D.J., Sawa, A.: Schizophrenia and epilepsy: is there a shared susceptibility? Neurosci. Res. **63**(4), 227–235 (2009)
8. Striano, P., Mancardi, M.M., Biancheri, R., Madia, F., Gennaro, E., Paravidino, R., Beccaria, F., Capovilla, G., Bernardina, B.D., Darra, F., et al.: Brain MRI findings in severe myoclonic epilepsy in infancy and genotype-phenotype correlations. Epilepsia **48**(6), 1092–1096 (2007)
9. Dabbs, K., Becker, T., Jones, J., Rutecki, P., Seidenberg, M., Hermann, B.: Brain structure and aging in chronic temporal lobe epilepsy. Epilepsia **53**(6), 1033–1043 (2012)
10. Elsayed, A., Coenen, F., Jiang, C., Garcia-Finana, M., Sluming, V.: Corpus callosum MR image classification. In: Proceedings AI 2009, pp. 333–348. Springer (2009)
11. Elsayed, A., Coenen, F., Jiang, C., Garcia-Finana, M., Sluming, V.: Corpus callosum MR image classification. Knowl.-Based Syst. **23**(4), 330–336 (2010)
12. Long, S., Holder, L.B.: Graph-based shape analysis for MRI classification. Int. J. Knowl. Discov. Bioinform. **2**(2), 19–33 (2011)
13. Rousson, M., Paragios, N., Deriche, R.: Implicit active shape models for 3-D segmentation for MR imaging. In: Medical Image Computing and Computer-Assisted Intervention, pp. 209–216 (2004)
14. Cootes, T., Taylor, C., Cooper, D., Graham, J.: Active shape models-their training and application. Comput. Vis. Image Underst. **61**(1), 38–59 (1995)
15. Osher, S., Sethian, J.A.: Fronts Propagating with curvature-dependent speed: algorithms based on Hamilton-Jacobi formulations. J. Comput. Phys. **79**(1), 12–49 (1988)
16. Kulworatit, C., Khaosumang, T., Ungcharoensuk, N., Udomchaiporn, A.: Segmentation of lateral ventricles from 3-D magnetic resonance imaging. In: The 9th ECTI Conference on Application Research and Development, July 2017
17. Udomchaiporn, A., Coenen, F., Sluming, V., Garca-Finana, M.: 3-D MRI brain scan feature classification using an oct-tree representation. In: Proceedings of the 9th International Conference on Advanced Data Mining and Applications, December 2013
18. Mardia, K., Hainsworth, T.: A spatial thresholding method for image segmentation. IEEE Trans. Pattern Anal. Mach. Intell. **10**(6), 919–927 (1988)
19. Kass, M., Witkin, A., Terzopoulos, D.: Snakes: active contour models. Int. J. Comput. Vis. **1**(4), 321–331 (1988)
20. Reza, A.M.: Realization of the contrast limited adaptive histogram equalization (CLAHE) for real-time image enhancement. J. VLSI Signal Process. Syst. Signal Image Video Technol. **38**(1), 35–44 (2004)
21. https://www.mathworks.com/products/matlab.html. Accessed 2 Feb 2019
22. Altman, D.G.: Practical Statistics for Medical Research. Chapman and Hall (1991)

Deep Convolutional Neural Network with Edge Feature for Image Denoising

Supakorn Chupraphawan and Chotirat Ann Ratanamahatana$^{(\boxtimes)}$

Department of Computer Engineering, Faculty of Engineering,
Chulalongkorn University, 254 Phayathai Road., Bangkok 10330, Thailand
6070325121@student.chula.ac.th,
chotirat.r@chula.ac.th

Abstract. Image denoising is a classical challenge in computer vision and has attracted a large amount of research in the past few decades in attempts to find new approaches to denoise various types of images. The endeavors have been even more striking with the recent inception of deep learning. Deep learning has a well-known strength in its accuracy and effectiveness but with the main drawback of very long converging time during deep learning network training. Moreover, current state-of-the-art denoising techniques still lack the edge feature, one of the crucial attributes for sharp images. Therefore, in this paper, we propose a novel Deep Convolutional neural network with Edge Feature (DCEF) for denoising Additive White Gaussian Noise (AWGN). However, as the deep learning model training takes too much time to converge, we also propose an adaptive learning rate using a triangle technique that allows much faster converging time comparing to state-of-the-art approaches. Our DCEF demonstrates that it outperforms existing state-of-the-art approaches in terms of average PSNR scores in $\sigma = 15$ and 25 by 0.2 and 0.3, respectively, while achieving high MS-SSIM scores and using much fewer iterations to converge.

Keywords: Image denoising · Convolutional neural network · Edge detection

1 Introduction

Image denoising has one major goal to recover a clean image from an image that is disrupted by some noise. Image denoising problem is a classical challenge in computer vision and has attracted a large amount of research in the past few decades in attempts to find new approaches to denoise various types of images. It is a low-level task that is crucial for other tasks such as image classification, image segmentation, image detection, etc. Various approaches have been proposed for image denoising including the state-of-the-art BM3D [1], NLM [2], and WNNM [3].

With the recent inception and rapid improvement of GPU computing and deep learning, the researcher's endeavors in applying them to various problems and fields have never been more striking, including computer vision problems. For example, Simonyan et al. [4] perform image classification tasks that bring out impressive performance of a very deep convolutional neural network; He et al. [5] show image object detection by Residual Network (Res-net) and other tasks, demonstrating the ability of

© Springer Nature Switzerland AG 2020
P. Boonyopakorn et al. (Eds.): IC2IT 2019, AISC 936, pp. 169–179, 2020.
https://doi.org/10.1007/978-3-030-19861-9_17

deep learning's computation that could achieve more effective results and higher performance with GPU's framework. For image denoising problems, Burger et al. [6] demonstrate that a plain MLP could compete with the-state-of-the-art image denoising approach (BM3D) by preparing a large training dataset and deep network layers and huge number of neurons. Schuler et al. [7] show another approach by using similar structure that is then applied to remove artifacts caused by image deconvolution, and Jain et al. [8] apply a CNN model integrated with a Markov Random Field (MRF) to denoise noisy images. Deep learning has a well-known strength in its accuracy and effectiveness but with the main drawback of very long converging time during deep learning network training.

However, regardless of its strength in accuracy and effectiveness, deep learning suffers from its inefficiency as it requires substantial amount of time until convergence during deep learning network training, especially in large training sets. Moreover, current state-of-the-art denoising techniques still lack the edge feature, one of the crucial features contributing for sharp images. Yim et al. [9] demonstrate that applying a CNN model to noisy images generally produces unfavorable effects on edge structure, while requiring a large amount of time for training large datasets due to slow convergence. In fact, this edge feature, which could be obtained through various edge detection algorithms, could help the deep learning network analyze and locate objects in an image.

Therefore, in this paper, we propose a novel Deep Convolutional neural network with Edge Feature (DCEF) for denoising Additive White Gaussian Noise (AWGN). First, to overcome the edge feature problem in a noisy image, we propose a deep convolutional neural network, which has an ability to combine edge feature with the input image. Second, to expedite convergence of the training time, we propose an adaptive learning rate approach to improve the learning rate while allowing fast converge. Finally, we further optimize a weight initialization of a Convolutional Neural Network.

The contributions of this paper are summarized as follows:

1. We propose a novel Deep Convolutional neural network with Edge Feature (DCEF) for image denoising for additive white Gaussian noise (AWGN).
2. The network uses only a few iterations to converge during the training step for a large training dataset by applying a Cyclical learning rate technique.

2 Related Work

2.1 Image Denoising

A major goal of the image denoising task is to estimate a clean image $\mathbf{x}$ from a noisy image $\mathbf{y}$ with a standard deviation $\boldsymbol{\sigma}$. As shown in an image degradation model Eq. (1), a noisy image y can be constructed from a clean image x with some additive noise v, such as Additive White Gaussian Noise (AWGN), Salt-and-Pepper noise (S&P), Shot noise, etc.

$$y = x + v \tag{1}$$

Since Salt-and-Pepper noise can usually be eliminated by a median filter [10], and Shot noise has the same distribution as the Additive White Gaussian Noise when the standard deviation gets higher, in this paper, we therefore will focus on AWGN which has an independent noise from a mean value and a standard deviation σ. Dabov et al. [1] propose an image denoising technique using sparse 3D transform-domain collaborative filtering (BM3D), which consists two steps, i.e., first, finding some similar patches and combining them to the 3D block, then denoising it using 3D transformation. Gu et al. [3] use the advantage of a low-rank matrix, which is the prior knowledge that a large singular value of the low-rank matrix represents a majority of an image (WNNM). However, these approaches have two main limitations. First, these models involve several handcrafted and specific features to achieve good performance. Second, they also face complex optimization problems during testing, which is quite time consuming. After an advent of deep learning, Zhang et al. [11] propose a deep convolutional neural network (DnCNN) using a deep CNN model with a residual learning and batch normalization for extracting a clean image from a noisy image. However, the approach is still quite time consuming to train the network for large datasets.

2.2 Cyclical Learning Rate

Researchers are well aware of the time constraint during the network training and fine-tuning steps of the deep learning approaches. Normally, hyper parameters, including the learning rate, need to be tuned, but it usually is a time-consuming process. If the learning rate parameter is set to be too large, its gradient descent may overshoot a minimum. But if the learning rate parameter is set to be too small, its gradient descent could be too slow, and the model may not converge fast enough. Smith et al. [12] therefore propose a new approach to adaptively tune the learning rate and show that it is capable to train the models on large images datasets, i.e., CI-FAR 10 and CI-FAR 100 [13]. However, it has been discovered that a cyclical learning rate technique, which changes the learning rate parameter at every iteration according to some cyclic function, allows the model to adjust neuron weights more flexibly than the static learning rate approach. Specifically, it allows the learning rate parameter to vary between reasonable bounding values, and each cycle has a fixed length in terms of the number of iterations [12]. An example of a cyclical learning rate is shown in Fig. 1.

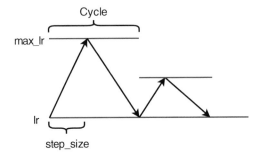

Fig. 1. An example of a cyclical learning rate policy. It has a starting point at 0th iteration then the learning rate increases at every iteration until reaching the maximum then decreases down to the minimum of 0, before repeating another cycle by reducing the maximum value by half in each subsequent cycle.

2.3   Peak Signal-to-Noise Ratio (PSNR)

Peak signal-to-noise ratio (PSNR) is one of the most well-known and standard metrics to evaluate the quality of the reconstructed image, comparing with the ground-truth image in terms of image overview. In image denoising task, this PSNR metric is used widely due to its fast evaluation process as the approach is based on a simple mean-squared error (MSE) value, as shown in Eqs. (2) and (3). Moreover, PSNR metric is comported for evaluating various image types. High PSNR means that the image has good quality and less error.

$$MSE = \sum_{M,N} [I_1(m,n) - I_2(m,n)]^2/M * N \qquad (2)$$

m and n are the width and the height of an image.

$$PSNR = 10 \log_{10}(MAX_I^2/MSE) \qquad (3)$$

MAX_I is a maximum possible pixel value of the image.

2.4   Multi-scale Structural Similarity (MS-SSIM)

Another metric for image denoising task is Multi-scale structural similarity (MS-SSIM) [14], which is more conscious on image structure than PSNR metric. Furthermore, MS-SSIM can handle different image scales and performs well on different subjective images and videos. MS-SSIM has an upper boundary score of 1 and lower boundary score of 0 that are used to measure the quality of an image. Equation (4) shows the MS-SSIM equation, which involves computations of luminance $l(x,y)^{\alpha}$, contrast $c(x,y)^{\beta}$

and structure $s(x, y)^\gamma$ between a ground-truth image x and a denoised image y, and α, β, γ are set to 1.

$$MS - SSIM(x, y) = \left[l(x, y)^\alpha * c(x, y)^\beta * s(x, y)^\gamma \right] \tag{4}$$

3 Proposed Method

In this section, we propose a new image denoising approach based on CNN model for additive white Gaussian noise (AWGN). An overview of the network architecture is described in Sect. 3.1. Our network architecture has two inputs, i.e., a noisy image and its edge feature image obtained by an edge detection algorithm that will be described in Sect. 3.2. The noisy image and its discovered edge feature are fed to the deep convolution neural network model, where feature extraction is done on primary layers, image feature processing is done on middle layers, and reconstruction process for estimating a denoised image is done in the final layer.

3.1 A Network Architecture

An overview of our proposed network is shown in Fig. 2. Our network is composed of several layers of Convolutional layers to extract image features, the Rectified Linear Unit layers (ReLU), and the Batch Normalization layers (BN), which will be described in this section.

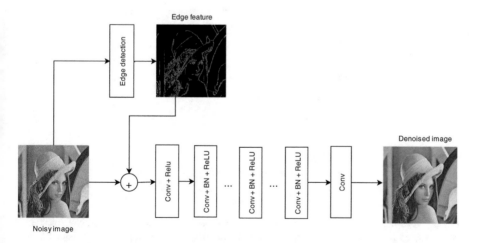

Fig. 2. An overview of our proposed network.

Input of our network is a noisy image and edge feature of the image and our goal is estimating a denoised image that closely the ground-truth image of noisy image. For our model, we use 16 layers in this network. In our first layer which is a Conv+ReLU activation function. The first Conv layer has 64 filters of size $3 \times 3 \times c$ to generate 64 feature maps, and Rectified Linear Units (ReLU) are used for nonlinearity activation function. c represents the number of image channels, e.g., $c = 3$ for a color image, and $c = 1$ for a greyscale image. Then Conv+BN+ReLU for 2nd to 15th layer has same structure as the first Conv layer. In the 16th layer (last layer), only a Conv layer with a filter size of $3 \times 3 \times 64$ is used for feature reconstruction, and finally produce a denoised image. With Conv+ReLU layers, our model removes an image structure from a noisy image gradually through hidden layers. For Batch Normalization layer (BN), this layer facilitates speed of model learning and supports model's performance.

3.2   An Edge Detection Algorithm

Edge feature for a model on image denoising task is needed because when a deep CNN model attempts to obtain some features of an image including edge feature which is disturbed by some noise, this feature may not useful for a deep CNN model. Diversely, if the model derives these features without or less noise, it will become useful. As a result, we propose a new approach for supporting the model that is using an edge feature from a Candy edge detection algorithm [15] that has a noisy image as an input. Candy edge detection algorithm is selected since it has been shown to outperform other edge detection algorithms [16]. With a Candy edge detection algorithm, the algorithm first removes noise, with a low pass filter, then uses non-maximum suppression to pick the best pixel for edge feature when there are multiple possibilities in a local neighborhood. Lastly, Candy edge detection algorithm extracts edge feature of a given noisy image.

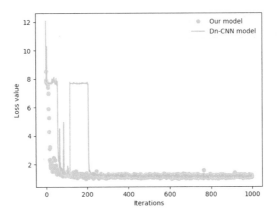

Fig. 3. A comparison of converging time between our model and DnCNN model. It demonstrates that our model converges in a shorter time using fewer iterations than DnCNN model when Cyclical learning rate technique is applied.

4   Experiments and Results

4.1   Experiment Setup

In this experiment, we use a standard publicly available dataset BSD400[1] [17] for model training; greyscale images with 180×180 resolution are used. We use TensorFlow's framework for our model's and DnCNN [10]'s implementation. We reimplement BM3D [1] using OpenCV, python, and C++ libraries with the same settings reported in the original papers.

For our model, we use Adam Optimizer [18] for optimization function and use Mean Squared Error between a denoised image and a ground-truth image for a loss function. The Cyclical learning rate in the first iteration is started at 0.003 and increased linearly to 0.005, after dropping down to the starting value. The subsequent iterations are repeated for 1,000 iterations, each time reducing the maximum value by half. We use Xavier initializer for initializing the weight parameters in every conv2d layer except for the last layer. We set an effective patch size as 40×40 and crop 128×1600 patches to train our model with the noise levels of 15, 25 and 50. These Sigma levels are used based on existing work where it is well recognized by the community that $\sigma = 15$ is considered a lightweight noise level, $\sigma = 25$ is considered a medium noise level, and $\sigma = 50$ is considered a heavy noise level for real environment. The model is trained with 50 epochs for all noise levels.

Fig. 4. Set12 [20] dataset for testing in this experiment

---

[1] https://github.com/cszn/DnCNN/tree/master/TrainingCodes/DnCNN_TrainingCodes_v1.0/data/Train400.

To evaluate our performance, we compare our model with state-of-the-art algorithms, i.e., BM3D [1] and DnCNN [11]. We use the same datasets, BSD68[2] [19] and Set12[3] [20], as in [11] for model testing. BSD68 dataset contains 68 natural images from Berkley segmentation dataset, and Set12 is another well-known dataset that has 12 natural images for image denoising, as shown in Fig. 4. Both PSNR and MS-SSIM are used as our evaluation metrics.

All experiments are tested on Intel® core™ i7-3770 @ 3.40 GHz and 24 GB of Ram with a GeForce GTX1080Ti on Elementary OS Loki 0.4.1 Loki (Based on Ubuntu 16.04).

4.2   Results and Experiments Analysis

In this section, we assess our model performance and compare it with state-of-the-art denoising models, i.e., BM3D and DnCNN.

Tables 1 and 2 show the PSNR and MS-SSIM scores of denoised images on BM3D, DnCNN, and our DCEF from Set12 and BSD68 datasets, respectively, which demonstrates that our model outperforms both BM3D and DnCNN in terms of average PSNR (dB) on $\sigma = 15$ and 25. Specifically, our model works well on "Starfish", which has irregular structure on all sigma levels, and could benefit more from the edge feature comparing to other baselines. Nevertheless, with $\sigma = 50$, our model does not do well in preserving image structures due to intensive noise level. On the other hand, BM3D achieves the highest score on "House" and "Barbara", which contain repetitive structures; they could benefit from non-local-similarity priors, the basis of BM3D algorithm. However, in Fig. 5, the denoised image from BM3D evidently has a poor result on blurred image and images with artifacts. Similarly, a denoised image from DnCNN contains noticeable black artifacts. On the other hand, our denoised image has relatively better result without black artifact.

Regarding the training time on DnCNN and our DCEF, Fig. 3 shows that our model starts with loss value less than DnCNN's loss value and decreases rapidly. On the other hand, DnCNN model starts with higher loss value and gradually drops off and fluctuates one period then gradually converges. It indicates that Cyclical learning rate policy has some impact of model learning and neuron's weights adjustment, yielding satisfactory results on $\sigma = 15$ and 25 for both BSD68 and Set12 datasets.

---

[2] https://github.com/cszn/DnCNN/tree/master/TrainingCodes/DnCNN_TrainingCodes_v1.0/data/Test/Set68.

[3] https://github.com/cszn/DnCNN/tree/master/TrainingCodes/DnCNN_TrainingCodes_v1.0/data/Test/Set12.

Table 1. PSNR (dB)/MS-SSIM of BM3D, DnCNN, and our DCEF on **Set12 dataset** at sigma levels of 15, 25 and 50. **The best PSNR results** are highlighted in **boldface**.

Method (σ = 15)	BM3D	DnCNN	DCEF
Monarch	31.85/0.886	32.097/0.911	**33.010**/0.965
Boat	32.13/0.770	**34.438**/0.889	32.213/0.886
Airplane	31.07/0.834	**32.884**/0.937	31.497/0.920
Starfish	31.14/0.802	31.987/0.940	**32.109**/0.941
House	**34.93**/0.846	32.684/0.959	34.711/0.894
Lena	34.26/0.846	31.412/0.916	**34.313**/0.913
Parrot	31.37/0.839	31.474/0.925	**32.260**/0.903
Camera man	31.91/0.831	**34.195**/0.910	32.362/0.919
Man	31.92/0.837	**32.170**/0.930	31.542/0.923
Barbara	**33.10**/0.849	32.127/0.886	32.416/0.933
Peppers	32.69/0.855	32.215/0.903	**33.105**/0.940
Couple	32.14/0.752	32.100/0.903	**32.293**/0.904
Average	31.070/0.872	32.482/0.917	**32.653**/0.920
Method (σ = 25)	BM3D	DnCNN	DCEF
Monarch	29.25/0.846	**30.247**/0.942	30.078/0.852
Boat	29.90/0.728	**30.083**/0.840	30.015/0.857
Airplane	28.42/0.771	29.026/0.894	**32.882**/0.872
Starfish	28.56/0.770	29.239/0.900	**30.700**/0.912
House	32.85/0.734	**32.914**/0.874	28.971/0.889
Lena	32.07/0.774	**32.285**/0.885	29.421/0.890
Parrot	28.93/0.771	30.002/0.852	**30.075**/0.839
Camera man	29.45/0.739	**30.070**/0.881	29.995/0.879
Man	29.61/0.775	29.419/0.891	**29.846**/0.895
Barbara	**30.71**/0.810	29.810/0.897	29.328/0.900
Peppers	30.16/0.811	**30.635**/0.909	30.391/0.941
Couple	29.71/0.711	29.922/0.857	**32.291**/0.884
Average	29.9690/0.770	30.3040/0.885	**30.3330**/0.884
Method (σ = 50)	BM3D	DnCNN	DCEF
Monarch	25.82/0.724	26.886/0.886	**26.895**/0.795
Boat	26.78/0.571	27.098/0.739	**29.748**/0.816
Airplane	25.10/0.623	25.983/0.826	**27.213**/0.843
Starfish	25.04/0.645	25.608/0.807	**25.692**/0.810
House	**29.69**/0.583	29.888/0.831	26.781/0.876
Lena	29.05/0.619	**29.399**/0.825	25.482/0.818
Parrot	25.90/0.633	**27.115**/0.743	26.027/0.813
Camera man	26.13/0.577	26.916/0.805	**28.577**/0.812
Man	**26.81**/0.633	26.494/0.824	26.079/0.792
Barbara	**27.22**/0.637	26.263/0.800	26.993/0.736
Peppers	26.68/0.675	**27.381**/0.847	26.749/0.737
Couple	26.46/0.569	**26.861**/0.754	26.771/0.752
Average	26.722/0.624	**27.158**/0.807	26.917/0.800

Table 2. **Average** PSNR (dB)/MS-SSIM for **BSD68 test dataset** on BM3D, DnCNN, and DCEF. **The best PSNR results** are highlighted in **boldface**.

Method	BM3D	DnCNN	DCEF
$\sigma = 15$	31.070/0.872	31.493/0.890	**31.496**/0.906
$\sigma = 25$	28.570/0.801	29.108/0.855	**29.141**/0.855
$\sigma = 50$	25.621/0.686	**26.188**/0.746	26.047/0.740

Fig. 5. An example of denoised images from "Set12" by BM3D, DnCNN, and our model (DCEF) on sigma = 15. Our model, DCEF, could better preserve structures of a denoised image than other baselines.

5 Conclusion

In this paper, we propose a Deep Convolutional neural network with Edge Feature (DCEF) to denoise an image containing additive white Gaussian noise (AWGN) on $\sigma = 15$, 25, and 50, respectively. Candy edge detection algorithm is used to extract edge features from a grayscale image. Cyclical learning rate technique is used to help in faster convergence, comparing to other baselines. The model outperforms other baselines in terms of PSNR (dB) and MS-SSIM metrics, especially on averaged sigma of 15 and 25.

References

1. Dabov, K., Foi, A., Katkovnik, V., Egiazarian, K.: Image denoising by sparse 3-D transform-domain collaborative filtering. **16**, 2080–2095 (2007)
2. Buades, A., Coll, B., Morel, J.-M.: A non-local algorithm for image denoising. In: IEEE Computer Society Conference on Computer Vision and Pattern Recognition, CVPR, pp. 60–65. IEEE (2005)

3. Gu, S., Zhang, L., Zuo, W., Feng, X.: Weighted nuclear norm minimization with application to image denoising. In: Proceedings of the IEEE Conference on Computer Vision and Pattern Recognition, pp. 2862–2869 (2014)
4. Simonyan, K., Zisserman, A.: Very deep convolutional networks for large-scale image recognition (2014)
5. He, K., Zhang, X., Ren, S., Sun, J.: Deep residual learning for image recognition. In: Proceedings of the IEEE Conference on Computer Vision and Pattern Recognition, pp. 770–778 (2014)
6. Burger, H.C., Schuler, C.J., Harmeling, S.: Image denoising: can plain neural networks compete with BM3D? In: IEEE Conference on Computer Vision and Pattern Recognition (CVPR), pp. 2392–2399. IEEE (2012)
7. Schuler, C.J., Christopher Burger, H., Harmeling, S., Scholkopf, B.: A machine learning approach for non-blind image deconvolution. In: Proceedings of the IEEE Conference on Computer Vision and Pattern Recognition, pp. 1067–1074 (2013)
8. Jain, V., Seung, S.: Natural image denoising with convolutional networks. In: Advances in Neural Information Processing Systems, pp. 769–776 (2009)
9. Yim, J., Sohn, K.-A.: Enhancing the Performance of Convolutional Neural Networks on Quality Degraded Datasets (2017)
10. Huang, T., Yang, G., Tang, G.: A fast two-dimensional median filtering algorithm. **27**, 13–18 (1979)
11. Zhang, K., Zuo, W., Chen, Y., Meng, D., Zhang, L.: Beyond a gaussian denoiser: residual learning of deep cnn for image denoising **26**, 3142–3155 (2017)
12. Smith, L.N.: Cyclical learning rates for training neural networks. In: IEEE Winter Conference on Applications of Computer Vision (WACV), pp. 464–472. IEEE (2017)
13. Krizhevsky, A., Hinton, G.: Learning Multiple Layers of Features from Tiny Images. Citeseer (2009)
14. Wang, Z., Simoncelli, E.P., Bovik, A.C.: Multiscale structural similarity for image quality assessment. In: 2003 The Thrity-Seventh Asilomar Conference on Signals, Systems & Computers, pp. 1398–1402. IEEE (2003)
15. John, C.: A computational approach to edge detection, 679–698 (1986)
16. Acharjya, P.P., Das, R., Ghoshal, D.: Technology: study and comparison of different edge detectors for image segmentation (2012)
17. Martin, D., Fowlkes, C., Tal, D., Malik, J.: A database of human segmented natural images and its application to evaluating segmentation algorithms and measuring ecological statistics. In: Proceedings of Eighth IEEE International Conference on Computer Vision, ICCV, pp. 416–423. IEEE (2001)
18. Kingma, D.P., Ba, J.: Adam: a method for stochastic optimization (2014)
19. Roth, S., Black, M.J.: Fields of experts: a framework for learning image priors. In: IEEE Computer Society Conference on Computer Vision and Pattern Recognition, CVPR, pp. 860–867. IEEE (2005)
20. The USC-SIPI Image Database (1977)

The Combination of Different Cell Sizes of HOG with KELM for Vehicle Detection

Natthariya Laopracha[✉]

Computer Science, Faculty of Informatics,
Mahasarakham University, Talat, Thailand
natthariya@gmail.com

Abstract. HOG has been developed successfully in many intelligent vehicle detection systems. HOG still has interesting problems that consist of (i) redundant features and (ii) ambiguous features (similarities between vehicles and non-vehicles), which problems have an effect on time computation and misclassification. The vertical direction of HOG method (V-HOG) and adding the position of orientation bins and intensity features (πHOG) improve the problems of HOG; but they produce redundant and ambiguous features in various regions of vehicles. This paper proposes a new method for improving the performance of HOG that has flexibility in various regions of vehicles. The proposed method used combines different sized cells of HOG that is called CDC-HOG. The CDC-HOG were conducted on a GTI dataset, which consists of four regions (far, front, left, and right regions). The CDC-HOG is compared with HOG, V-HOG, πHOG, and PHOG; uses the kernel extreme learning machine (KELM), and supports vector machine (SVM) for evaluating features. The CDC-HOG with KELM produced the highest performance in terms of accuracy, true positive rate, and false positive rate for all regions.

Keywords: Vehicle detection · Feature extraction ·
Histograms Oriented Gradients (HOG) · Support Vector Machine (SVM) ·
Kernel Extreme Learning Machine (KELM)

1 Introduction

Recently, intelligent vehicle detection systems [1] have been applied in various applications; for example, automatic cars [2], traffic surveillance [3], application for finding a car park [4], and counting vehicles. Most vehicle detection systems developed are vision-based because it is easier to understand than sensor-based. The vision-based system consists of two steps [5] which are hypothesis generation (HG) and hypothesis verification (HV). The HG has three basic categories of methods: (1) knowledge, (2) stereo-vision-based, and (3) motion-based. In a complex environment, using only the HG step may result in the inability to classify objects as vehicles or non-vehicles. The HV method can classify objects in a complex environment [6]. HV uses the feature-extraction method and machine learning for classifying vehicles or non-vehicles. Histograms of oriented gradients (HOG) [7] is popular for extracting features in vehicle detection because it is robust in varied light conditions, in complex

P. Boonyopakorn et al. (Eds.): IC2IT 2019, AISC 936, pp. 180–190, 2020.
https://doi.org/10.1007/978-3-030-19861-9_18

backgrounds, in various views of vehicles, and in multi-angle views of vehicles. There are many successful research applications of HOG for vehicle detection systems. HOG with Support Vector Machine (SVM) is developed in traffic flow applications, which can describe vehicles and non-vehicles in dense traffic flow [8], and are applied for estimating inter-vehicle distances [9]. HOG combines with Harr and Adaboost in the Monocular system, which alerts a driver about a possible collision with other vehicles in various conditions [10].

More recently, deep learning has produced a high detection rate for vehicle detection; however, deep learning has a disadvantage [11] in vehicle detection. First, deep learning requires a large number of samples for training [4]. Second, the structure of deep learning needs to find a suitable dataset, and different datasets may not be compatible with other structures. Lastly, the input layer of deep learning is varied in terms of rotation and variations images. This problem may be related to preparing images before feeding to deep learning by using HOG [12], which can increase the accuracy of deep learning.

HOG has produced high performance in vehicle detection, which can apply in machine learning and deep learning. However, HOG consists of redundant features [13] and ambiguous features [14] that have an effect on long computation times and misclassification. Although there are vertical direction methods of HOG (V-HOG) [13], such as adding the position of orientation bins and intensity features (πHOG) [14], and pyramid HOG (PHOG) [15] for solving these problems; these methods are unable to increase the performance of HOG in various regions of vehicles.

This paper proposes a new method for extracting features that are able to describe features of vehicles in various regions, which methods are clearer than HOG, V-HOG, πHOG, and PHOG. The proposed method combines two different sized cells of HOG (CDC-HOG) by using the big size cell of HOG, which can reduce redundant features and can more clearly describe the difference between vehicles and non-vehicles. The combination can describe ambiguous features between vehicles and non-vehicles. Therefore, the proposed method can produce the performance in terms of accuracy, true positive rate, and false positive rate better than HOG, V-HOG, πHOG, and PHOG in various regions of vehicles.

2   The Proposed Combination of Different Size Cells of HOG (CDC-HOG)

Dalal et al. [16] proposed a HOG descriptor for representing the human image. The HOG produced high performance in low-resolution images, in varied light conditions, and complexity of backgrounds. Therefore, HOG is popular in several systems such as in vehicle detection, human detection, and president detection. The HOG divides the images into many cells. The size of cells (s) have an effect on the accuracy and time computation of HOG, and can be assigned dependents by the user; for example 4, 8, 16, and 32 pixels when fixing the size of the image as 64×64 pixels (see in Fig. 1). The HOG uses a block for sliding from left to right and above-to-down. Figure 2 presents the 1st block (red color) and 2nd block (green color) that appear to be overlapping block when there is a sliding block. The overlapping block produces redundancy features and ambiguous features (features similar between vehicles and non-vehicles). The small size of cell produces the overlapping number more than the big size of the cell. For example, $s = 4$ produced the number of overlapping 224 blocks, $s = 8$ produced the number of overlapping 48 blocks, $s = 16$ produced the number of overlapping 8 blocks, and $s = 32$ produced the number of overlapping 0 block. Therefore, $s = 4$ and $s = 8$ produced the number of redundant features and ambiguous features more than $s = 16$ and $s = 32$. The cell size of 32 is interested in extracting features. The πHOG uses $s = 32$ for improving the performance of HOG and adding the position of orientation bins and intensity features into HOG features that are called PIHOG or πHOG [4]. However, some position of orientation bins of πHOG produced similarities between vehicles and non-vehicles for various regions, which is a disadvantage of πHOG. Moreover, the computation position of orientation bins of πHOG produced long computation times that do not support real time.

V-HOG reduces redundant features of HOG by using only the vertical direction of HOG. The V-HOG can reduce the redundant features and time computation of HOG. However, using only the vertical direction results in inflexibility in various regions of vehicles.

This paper uses $s = 32$ for reducing the redundant features of HOG from overlapping blocks. However, $s = 32$ is unable to describe vehicles in various regions of vehicles. Thus, one uses $s = 16$ for combining features with $s = 32$ because $s = 16$ produces redundant features less than $s = 4$ and $s = 8$.

Figure 3 presents steps in the combination of different sized cells of 16 and 32 pixels. Using $s = 16$ consists of 9 blocks for sliding on an image and produces the number of overlapping 8 blocks; while $s = 32$ has one block on an image and it does not have overlapping blocks. The features of $s = 16$ (F1) are combined with features of $s = 32$, which features of CDC-HOG are collected in {F1, F2}. These features are classified by KELM or SVM.

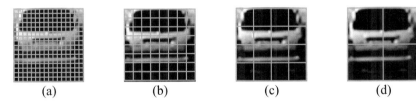

(a)   (b)   (c)   (d)

Fig. 1. Traditional division of cell sizes of HOG by using (a) $s = 4$, (b) $s = 8$, (c) $s = 16$, and (d) $s = 32$.

Ovelapping

Fig. 2. The overlapping block of HOG in $s = 8$.

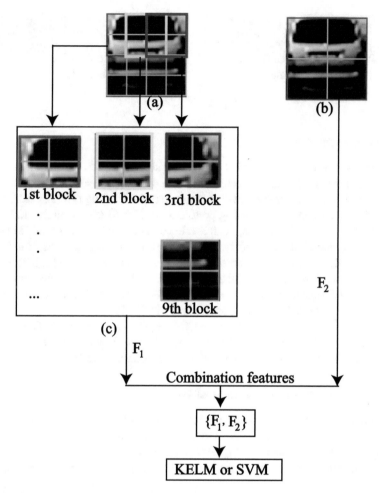

Fig. 3. The proposed method of combining different sizes of cells by using (a) $s = 16$ and (b) $s = 32$, (c) HOG extraction of $s = 16$, and (d) the result of combination features.

3 Experiment and Analysis

3.1 Dataset Preparation

The proposed method is conducted on a GTI public dataset [17]. The GTI includes only the rear of vehicles, which is divided into four regions: i.e., far region, front region, left region and right region. The far, left, and right regions contained 975 images for vehicles and 975 images for non-vehicles. The front region contained 500 images for vehicles and 975 images for non-vehicles. The size of the images is 64 × 64 pixels. The GTI dataset is shown in Fig. 4.

(a) vehicle images

(b) non-vehicle images

Fig. 4. GTI dataset

3.2   Experiment Setup

The parameters of the CDC-HOG consist of the size of the cell (s) and the number of orientation bins (β). The sizes of cells are assigned as $s = 4$, $s = 8$, $s = 16$, and $s = 32$. The β is fixed as 9. The experiment uses kernel extreme learning machine (KELM) [18] and SVM as classifiers for evaluating performance. KELM uses a single hidden layer, which is a fast method for learning features. In addition, the number of hidden nodes of KELM are produced automatically, which makes it easy to create the structure of a network. The KELM uses RBF kernel, and SVM uses a linear kernel [18]. The number of train sets and test sets includes 30 sets and each set divided in the train; thus tested as 50:50. The performance evaluation criteria use three values that are accuracy (acc.), false positive rate (FPR) and true positive rate (TPR) [19].

This study compares the performance CDC-HOG with HOG, V-HOG, πHOG, and PHOG. The size of the cell and β of HOG, V-HOG and πHOG use the same sizes with CDC-HOG. The PHOG sets the number of orientation bins as $\beta = 9$, 12, and 18.

3.3   Experiment Result of Traditional HOG

This section presents the experiment result of HOG in different sizes of cells within far, front, left, and right regions. Table 1 shows the performance of HOG by using KELM as a classifier. HOG produced the highest accuracy by using $s = 16$ in the far, front, and left region; while HOG produced the highest accuracy in the right region when using

Table 1. Performance of HOG within each region by using the KELM classifier.

s	Far			Front		
	Acc	TPR	FPR	Acc	TPR	FPR
4	97.32	0.9766	0.0300	96.50	0.9958	0.0485
8	98.20	0.9769	0.0126	98.11	0.9913	0.0238
16	**98.29**	0.9793	0.0134	**99.03**	0.9790	0.0038
32	96.77	0.9619	0.0262	98.85	0.9715	0.0024
s	Left			Right		
	Acc	TPR	FPR	Acc	TPR	FPR
4	96.10	0.9979	0.0706	92.44	0.9895	0.1252
8	98.20	0.9956	0.0307	96.66	0.9921	0.0563
16	**99.06**	0.9858	0.0044	98.32	0.9727	0.0059
32	98.44	0.9810	0.0121	**98.51**	0.9797	0.0093

$s = 32$. Table 2 shows the performance of HOG by using SVM as a classifier. HOG produced the highest accuracy by using $s = 4$ in far, front and right regions; while, the left region uses $s = 16$ that produced the highest accuracy. HOG produced the highest performance in far, left, and right regions when using KELM as a classifier. Meanwhile, HOG with SVM produced the highest performance in the front region.

Table 2. Performance of HOG within each region by using the SVM classifier.

s	Far			Front		
	Acc	TPR	FPR	Acc	TPR	FPR
4	**98.20**	0.9849	0.0209	**99.52**	0.9947	0.0045
8	98.04	0.9846	0.0238	99.48	0.9907	0.0031
16	98.19	0.9832	0.0192	99.07	0.9824	0.0049
32	95.05	0.9431	0.0416	98.52	0.9719	0.0077
s	Left			Right		
	Acc	TPR	FPR	Acc	TPR	FPR
4	98.87	0.9917	0.0142	**98.38**	0.9822	0.0145
8	98.78	0.9892	0.0135	98.14	0.9821	0.0193
16	**98.91**	0.9899	0.0117	97.83	0.9777	0.0209
32	98.11	0.9777	0.0152	97.86	0.9761	0.0187

3.4   Experiment Result of the Proposed Method

This section presents the experimental results of the proposed CDC-HOG method. The objective of the proposed method combines the sizes of cells of 16 with 32, or cell sizes 32 with 16. However, it shows all possible combination within $s = 4, 8, 16$, and 32 that consist of 12 combinations. Tables 3, 4, 5, and 6 show the performance of CDC-HOG by using KELM and SVM as classifiers within far, front, left, and right regions respectively. The combination of cell sizes of $s = 32$ and $s = 8$ produced the highest accuracy when using KELM as a classifier in far and front regions; however, combination cell sizes of $s = 16$ and $s = 32$ or $s = 16$ and $s = 32$ produced high performance when using KELM and SVM as classifiers (Tables 1 and 2). Within the left region, CDC-HOG produced the highest accuracy when the combined cell sizes of $s = 16$ and $s = 4$ used KELM as a classifier. However, the combination of $s = 16$ and $s = 32$ or $s = 32$ and $s = 16$ produced high performance when using KELM and SVM as classifiers (Table 5). Table 6 shows the performance in the right region. CDC-HOG produced the highest accuracy when the combination of cell sizes of $s = 32$ and 15 used KELM as a classifier. Therefore, the combination cell sizes of $s = 16$ and $s = 32$ or $s = 32$ and $s = 16$ produced high accuracy in all regions when using KELM as a classifier.

Table 3. Performance of CDC-HOG in different combination of cell sizes by using KELM and SVM as classifiers in the far region.

s	CDC-HOG with KELM			CDC-HOG with SVM		
	Acc	TPR	FPR	Acc	TPR	FPR
4_8	97.35	0.97654	0.029400	98.28	0.9858	0.0200
4_16	97.32	0.97665	0.030054	98.24	0.9846	0.0198
4_32	97.32	0.97652	0.029926	98.22	0.9850	0.0204
8_4	98.27	0.98074	0.015231	98.28	0.9858	0.0200
8_16	98.26	0.97754	0.012154	98.14	0.9849	0.0219
8_32	98.20	0.97668	0.012436	98.04	0.9842	0.0233
16_32	98.53	0.9798	0.009043	98.24	0.9846	0.0198
16_4	98.54	0.98086	0.009864	98.14	0.9849	0.0219
16_8	98.28	0.97879	0.013021	98.21	0.9834	0.0190
32_4	97.81	0.97394	0.017577	98.22	0.9850	0.0204
32_8	98.63	0.98547	0.012803	98.04	0.9842	0.0233
32_16	98.53	0.98585	0.015278	98.21	0.9834	0.0190

Table 4. Performance of CDC-HOG in different combinations of sizes of cells by using KELM and SVM as classifiers in the front region.

s	CDC-HOG with KELM			CDC-HOG with SVM		
	Acc	TPR	FPR	Acc	TPR	FPR
4_8	96.60	0.9956	0.0471	99.63	0.9955	0.0033
4_16	96.56	0.9956	0.0476	99.60	0.9955	0.0037
4_32	96.50	0.9958	0.0485	99.55	0.9948	0.0042
8_4	98.40	0.9937	0.0206	99.63	0.9955	0.0033
8_16	98.24	0.9916	0.0221	99.53	0.9913	0.0026
8_32	98.14	0.9913	0.0234	99.45	0.9903	0.0033
16_32	99.24	0.9873	0.0050	99.60	0.9955	0.0037
16_4	99.22	0.9832	0.0031	99.53	0.9913	0.0026
16_8	99.06	0.9794	0.0036	99.07	0.9830	0.0052
32_4	99.40	0.9936	0.0057	99.55	0.9948	0.0042
32_8	99.73	0.9976	0.0029	99.45	0.9903	0.0033
32_16	99.69	0.9948	0.0020	99.07	0.9830	0.0052

3.5 Comparison Performance of the Proposed Method with the Highest Performance of HOG, V-HOG, πHOG, and PHOG

This section summarizes the performance of CDC-HOG for comparing it with the highest accuracy of HOG, V-HOG, πHOG, and PHOG. The comparison performance is shown in Table 7. There are six values for evaluating that consist of accuracy (Acc),

Table 5. Performance of CDC-HOG in different combinations of sizes of cells by using KELM and SVM as classifiers in the left region.

s	CDC-HOG with KELM			CDC-HOG with SVM		
	Acc	TPR	FPR	Acc	TPR	FPR
4_8	96.19	0.9979	0.0690	98.97	0.9918	0.0124
4_16	96.13	0.9979	0.0700	98.90	0.9907	0.0126
4_32	96.11	0.9979	0.0705	98.90	0.9917	0.0136
8_4	98.29	0.9968	0.0302	98.97	0.9918	0.0124
8_16	98.28	0.9957	0.0293	98.85	0.9889	0.0119
8_32	98.23	0.9955	0.0301	98.81	0.9893	0.0130
16_32	99.07	0.9862	0.0047	98.90	0.9907	0.0126
16_4	99.11	0.9851	0.0027	98.85	0.9889	0.0119
16_8	99.07	0.9851	0.0036	99.01	0.9907	0.0104
32_4	98.31	0.9808	0.0145	98.90	0.9917	0.0136
32_8	98.74	0.9864	0.0114	98.81	0.9893	0.0130
32_16	98.92	0.9868	0.0084	99.01	0.9907	0.0104

Table 6. Performance of CDC-HOG in different combinations of sizes of cells by using KELM and SVM as classifiers in the right region.

s	CDC-HOG with KELM			CDC-HOG with SVM		
	Acc	TPR	FPR	Acc	TPR	FPR
4_8	92.66	0.9897	0.1220	98.46	0.9834	0.0142
4_16	92.55	0.9895	0.1234	98.58	0.9833	0.0116
4_32	92.46	0.9894	0.1248	98.42	0.9824	0.0139
8_4	96.95	0.9940	0.0526	98.46	0.9834	0.0142
8_16	96.93	0.9927	0.0519	98.36	0.9830	0.0157
8_32	96.77	0.9923	0.0544	98.27	0.9834	0.0179
16_32	98.46	0.9748	0.0052	98.58	0.9833	0.0116
16_4	98.36	0.9720	0.0043	98.36	0.9830	0.0157
16_8	98.37	0.9733	0.0054	97.99	0.9793	0.0194
32_4	98.00	0.9695	0.0089	98.42	0.9824	0.0139
32_8	98.57	0.9791	0.0075	98.27	0.9834	0.0179
32_16	98.83	0.9842	0.0075	97.99	0.9793	0.0194

True positive rate (TPR), False positive rate (FPR), the number of features (#Features), testing time (TS), and training time (TT).

The results in Table 7 can be summarized as follows:

(1) CDC-HOG with KELM produced the highest performance in terms of accuracy, TPR, and FPR for all regions.

Table 7. Comparison performance of CDC-HOG for each region with the highest accuracy of HOG, V-HOG, πHOG, and PHOG.

#	Method	Acc	TPR	FPR	#Feature	TS	TT	s	Classifier
Far									
1	CDC-HOG	98.63	0.9855	0.0128	1800	0.1231	0.1522	32_8	KELM
2	CDC-HOG	98.53	0.9798	0.0090	360	0.0412	0.0541	16_32	KELM
3	V-HOG	98.40	0.9831	0.0151	252	0.0185	0.0291	8	KELM
4	HOG	98.29	0.9793	0.0134	324	0.0357	0.0425	16	KELM
5	πHOG	98.11	0.9839	0.0161	616	0.0182	0.2401	16	SVM
6	PHOG	94.56	0.9204	0.0253	1530	0.1924	0.3562	-	SVM
Front									
1	CDC-HOG	99.73	0.9976	0.0029	1800	0.0295	0.0462	32_8	KELM
2	CDC-HOG	99.69	0.9948	0.0020	360	0.0091	0.0112	32_16	KELM
3	V-HOG	99.58	0.9973	0.0050	252	0.0032	0.0083	8	KELM
4	HOG	99.52	0.9947	0.0045	8100	0.1456	0.7171	4	SVM
5	πHOG	99.08	0.9936	0.0064	616	0.0061	0.0384	16	SVM
6	PHOG	95.22	0.9027	0.0195	1530	0.0263	0.1107	-	SVM
Left									
1	CDC-HOG	99.11	0.9851	0.0027	8424	0.7712	1.3041	16_4	KELM
2	CDC-HOG	99.07	0.9862	0.0047	360	0.0394	0.0512	16_32	KELM
3	HOG	99.06	0.9858	0.0044	324	0.0323	0.0401	16	KELM
4	πHOG	99.04	0.9914	0.0086	616	0.0127	0.1092	16	SVM
5	V-HOG	98.42	0.9838	0.0153	252	0.0164	0.0262	8	KELM
6	PHOG	97.01	0.9571	0.0159	1530	0.0456	0.1421	-	SVM
Right									
1	CDC-HOG	98.83	0.9842	0.0075	360	0.0351	0.0492	32_16	KELM
2	HOG	98.51	0.9797	0.0093	36	0.0001	0.0002	32	KELM
3	V-HOG	98.22	0.9756	0.0130	252	0.0158	0.0231	8	KELM
4	πHOG	98.15	0.9800	0.0168	2920	0.1332	0.5183	8	SVM
5	PHOG	95.95	0.9416	0.0210	1530	0.0352	0.1511	-	SVM

(2) V-HOG can increase the accuracy of TPR of HOG in far and front regions; but the V-HOG is unable to increase performance (accuracy, TPR, and FPR) in left and right regions. Also, the far and front regions have a similar rectangle, of which V-HOG uses only the vertical direction which produces high performance in these regions. Meanwhile, the left and right regions have multi-angles, which is the disadvantage of V-HOG.

(3) πHOG produced less performance than CDC-HOG and HOG in all regions because of some positions of orientation bin features, or some intensity features of πHOG consisting of ambiguous features. The ambiguous features affect the misclassification vehicles as non-vehicles, or non-vehicles as vehicles.

(4) PHOG produced a lower performance than CDC-HOG, HOG, V-HOG, and πHOG.

(5) CDC-HOG, using $s = 16$ with $s = 32$ or $s = 32$ with $s = 16$, can increase performance in terms of accuracy, TPR, and FPR for far, front, and right regions.

Meanwhile, CDC-HOG, using $s = 16$ with $s = 32$, can increase the accuracy and TPR of HOG for the left region.

(6) V-HOG produced a number of features shorter than HOG and faster than HOG in all regions; but V-HOG was unable to increase the performance in terms of accuracy, TPR, and FPR of HOG in all regions. While CDC-HOG, using $s = 16$ with $s = 32$ or $s = 32$ or $s = 16$, can increase performance of HOG in all regions, the CDC-HOG still results in faster testing times and training times.

The proposed CDC-HOG is a robust method for extracting features in various regions and has produced a compact number of features. Therefore, CDC-HOG is suitable in real time for vehicle detection.

4 Conclusion

This paper focuses on improving the performance of HOG by a combination of different cell sizes of HOG which is called CDC-HOG. The interesting problems of HOG are the redundant and ambiguous features, which problems affect time computation and misclassification. There are several methods for improving the performance of HOG by solving these problems. The V-HOG uses only the vertical direction for reducing redundant features. The πHOG adds the position of orientation bins and intensity features for describing ambiguous features between vehicles and non-vehicles. The V-HOG can increase the performance of HOG in far and front regions. The πHOG produced performance higher than V-HOG in the left region but it produced lower performance than HOG in all regions. The V-HOG and πHOG were meant to increase the performance of HOG in various regions. Meanwhile, CDC-HOG used the size of cell, as did πHOG which is $s = 32$ for reducing redundant features. Then, the CDC-HOG uses the combination features of $s = 32$ with $s = 16$ for describing ambiguous features between vehicles and non-vehicles. While πHOG uses the position of orientation bins and intensity for describing the difference between vehicles and non-vehicles features; the CDC-HOG produced the performance in terms of accuracy, TPR, and FPR better than πHOG in all regions. In addition, CDC-HOG produced a number of features shorter than πHOG in all regions. Lastly, CDC-HOG produced the best performance for all regions when compared with HOG, V-HOG, πHOG, and PHOG, as well as produced compact vector features, which can result in fast computation in real time.

In future work, this study plans to develop CDC-HOG for deep learning and apply objective classification; for example, for pedestrians, animals, sign traffic, and etc.

References

1. Liu, S., Wang, S., Shi, W., Liu, H., Li, Z., Mao, T.: Vehicle tracking by detection in UAV aerial video. Sci. China Inf. Sci. **62**(February), 1–3 (2019)
2. Dai, X.: Signal processing: image communication HybridNet : a fast vehicle detection system for autonomous driving. Signal Process. Image Commun. **70**(September), 79–88 (2019)

3. Tian, B., et al.: Hierarchical and networked vehicle surveillance in ITS: a survey. IEEE Trans. Intell. Transp. Syst. **18**(1), 25–48 (2017)
4. De Almeida, P.R., Oliveira, L.S., Britto Jr., A.S., Silva Jr., E.J., Koerich, A.L.: PKLot–a robust dataset for parking lot classification. Expert Syst. Appl. **42**(11), 4937–4949 (2015)
5. Zaarane, A., Slimani, I., Hamdoun, A., Atouf, I.: Real-time vehicle detection using cross-correlation and 2D-DWT for feature extraction, vol. 2019 (2019)
6. Wen, X., Shao, L., Fang, W., Xue, Y.: Efficient feature selection and classification for vehicle detection. IEEE Trans. Circ. Syst. Video Technol. **25**(3), 508–517 (2015)
7. Wei, Y., Tian, Q., Guo, J., Huang, W., Cao, J.: Multi-vehicle detection algorithm through combining Harr and HOG features. Math. Comput. Simul. **155**, 130–145 (2018)
8. Huy, V.P., Lee, B.-R.: Front-view car detection and counting with occlusion in dense traffic flow. Int. J. Control Autom. Syst. **13**(5), 1150–1160 (2015)
9. Huang, D.Y., Chen, C.H., Chen, T.Y., Hu, W.C., Feng, K.W.: Vehicle detection and inter-vehicle distance estimation using single-lens video camera on urban/suburb roads. J. Vis. Commun. Image Represent. **46**, 250–259 (2017)
10. Wang, X., Tang, J., Niu, J., Zhao, X.: Vision-based two-step brake detection method for vehicle collision avoidance. Neurocomputing **173**, 450–461 (2016)
11. Li, Y., Zhang, H., Xue, X., Jiang, Y., Shen, Q.: Deep learning for remote sensing image classification: a survey. Wiley Interdiscip. Rev. Data Min. Knowl. Discov. **8**, 1–17 (2018)
12. Gao, Y., Jong Lee, H.: Local tiled deep networks for recognition of vehicle make and model. Sensors **16**(2), 1–13 (2016)
13. Arrospide, J., Salgado, L., Camplani, M.: Image-based on-road vehicle detection using cost-effective histograms of oriented gradients. J. Vis. Commun. Image Represent. **24**(7), 1182–1190 (2013)
14. Kim, J., Baek, J., Kim, E.: A novel on-road vehicle detection method using π HOG. IEEE Trans. Intell. Transp. Syst. **16**(6), 3414–3429 (2015)
15. Bai, Y., Guo, L., Jin, L., Huang, Q.: A novel feature extraction method using pyramid histogram of orientation gradients for smile recognition. In: Proceedings of International Conference on Image Processing, ICIP, no. 07118074, pp. 3305–3308 (2009)
16. Dalai, N., Triggs, B.: Histograms of oriented gradients for human detection. In: Proceeding of the IEEE Conference on Computer Vision and Pattern Recognition, pp. 886–893 (2005)
17. The GTI-UPM Vehicle Image Database (2012). http://www.gti.ssr.upm.es/research/gti-data/databases
18. Fu, H., Vong, C.-M., Wong, P.-K., Yang, Z.: Fast detection of impact location using kernel extreme learning machine. Neural Comput. Appl. **27**(1), 121–130 (2016)
19. Ji, H., Wang, Y., Qin, H., Wang, Y., Li, H.: Comparative performance evaluation of intrusion detection methods for in-vehicle networks. IEEE Access **6**, 1 (2018)

Network, Cloud and Management

Dynamic Data Management for an Associative P2P Memory

Supaporn Simcharoen[✉]

Faculty of Mathematics and Computer Science,
FernUniversität in Hagen, Hagen, Germany
supaporn.simcharoen@gmail.com

Abstract. Dynamic Hash Tables (DHT) are efficient tools which help to improve the system's performance and reduce the overhead of search in peer-to-peer (P2P) systems. Chord exhibits significant disadvantages, such as keeping the node at a fixed position and storing entries on a fixed member peer. Thus, load imbalances may occur. To overcome this situation, a new ring-like associative memory has been introduced. Unlike the Chord, a fully decentralized token-game is presented for which items are in a lexical ordered ring-list, instead of being fixed to a determined peer. However, the items may have been shifted along the ring. According to the obtained experimental results the token-game worked with an adaption of the number of peers on the ring-list without any global instances. Furthermore, a suitable load-balancing basing on the physical analogue of communicating tubes which has been derived from the current flow theory that suggested a simultaneous and idiosyncratic flow of water that can be applied to the flow of information.

Keywords: Associative memory · Load balancing · Peer-to-peer

1 Introduction and Motivation

Decentralized system or Peer-to-peer system (P2P, [1]), is presented as a possible alternative of the centralized client-server systems. The system is capable of managing enormous volume of data on the network. Efficient data searching on P2P is a non-trivial task i.e. for which broadcast is often used for searching and generating numerous messages. However, broadcasting causes; high volume network traffic, and utilizing more memory in case that peers are required for all search operation [2]. Therefore, Dynamic Hash Tables (DHT, [3]) are perceived to be an alternative and efficient tools that help to improve the performance and to reduce the overhead of search in P2P systems. This could be achieved by building regular searchable structures. Search expenditure is ln (N) which upon applying DHT, overhead of search in P2P systems can be minimized.

Chord [4], a distributed lookup protocol, is the most frequently used of the mapping function in DHT. A consistent hashing [4, 5] has been assigned keys to Chord node (i.e., SHA-1 [6]). On the other hand, Node has some identifiers *IDs* (hash of IP Address) that are ordered on the Chord ring and resource which are equipped with a

P. Boonyopakorn et al. (Eds.): IC2IT 2019, AISC 936, pp. 193–204, 2020.
https://doi.org/10.1007/978-3-030-19861-9_19

number of *keys* (hash of file name). The mapping method is used not only when new entries are inserted, but also when users need to find the data items [7].

Figure 1(a) shows an example of the identifier circle, the Chord ring with m = 6 which has ten nodes and stores five keys. The successor of identifier 10 is node 14, thus, key 10 would be placed at node 14, likewise, keys 24 and key 30 would be placed at node 32, key 38 and key 54 would be placed at node 38 and node 56 respectively. Figure 1(b) shows the finger table of node 8 (blue line). The first finger of node 8 point to node 14((8 + 20) mod 26). Similarly, the last finger of node 8 points to node 42 ((8 + 25) mod 26). Moreover, in Fig. 1b) an example of node 8 lookup key 54 (red line) are also presented. First of all, the largest finger of node 8 that precedes 54 is node 42, then node 8 will necessitate node 42 to resolve the query. Subsequently, node 42 will determine the largest finger in its finger table that precedes 54, i.e. node 51. Lastly, node 51 will discover that its own successor, node 56, succeeds key 54, and then to return node 56 to node 8.

□ is a key   ● is a node

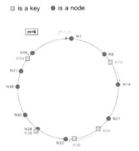 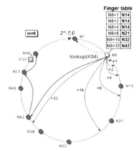

(a) An identifier circle consisting of ten nodes, with five keys to insert.

(b) The finger table of node 8 (blue line) and the path of node 8 lookup for key 54 (red line).

Fig. 1. A Chord ring with m = 6

As Chord has a set of significant disadvantages;

1. Entries are stored on a fixed member peer of the ring which is determined by the obtained hash value of their keys.
2. The position of a peer on the ring is fixed by the respective hash value of its IP address.

Hence, load imbalances may appear, for instance, if many entries are related to one and the same key or a small set of keys only, those peers may become overfull if users access a high number of one and the same data items. (This eventuation follows Zipf's law, which indicates that 10% of all contents are addressed by 90% of all search queries [8]).

To solve this problem, a new ring-like associative memory which is different from Chord has been introduced in the next section. Normally, peers in Chord are added when appear, then a fully decentralised token game will be decided when a new peer is needed. Moreover, the kept list is consisted of a lexical ordered ring-list, which allows items to move and unfixed to a determined peer. Furthermore, to attain a good speed and performance, the load distribution strategies and techniques should be added.

2 Modified Associated Ring Architecture

2.1 Concept

Associative memory is a set of storage locations that memory is addressed, depending on the content rather than the fixed address. A new ring-like associative memory structure of peer starts with a single peer. When entries are inserted, a lexical ordered ring-list of items is managed, each entry represents one HTML document for which a pair consisted of a key and a link ({<key>, <URL>} where the key is the centroid of its text information). Keys are words with a lexical order, i.e. that can be organized in a list. The memory shall be built; with a structure that is different from that of the ring which does not gathers every available peer of the Chord. Thus, the size of this ring will be changed depending on the number of entries that has been added. A decentral algorithm uses *the token game*, on the basis of random walkers, to run around the ring for updating the stage of the system.

When the ring is stable, providing that no more change in given items of every peer to its neighbor is found. In general, accessing to data that stored in memory can be simultaneously requested 90% of the time and related to the same topic. Consequently, load imbalance may occur which can be avoided by introducing the load balancing system.

The load balancing may become active when tokens pertain a high value while certain peers exhibits large imbalance which is built then a new peer is added. If all entries are in a lexical order ring-lists, any entries can be freely moved around the ring. These entries can locate their position fast and do not destroy the lexical order. For example, see Fig. 2.

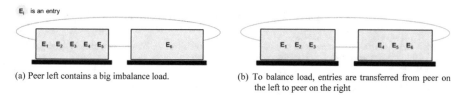

(a) Peer left contains a big imbalance load.   (b) To balance load, entries are transferred from peer on the left to peer on the right

Fig. 2.   To balance load with two peers

2.2 The Token Game

Decentral algorithm uses a token (T_i) to represent a free place on any peer of the ring that has been taken. The tokens run around the ring-list that are used for control addition and deletion of peers. A ring is stable if no token (or only one token) is found on the whole ring. As the global state is being considered in the absence of a global view, the token game will be required, i.e. allowed for random walkers to enter through the entire ring. If tokens are discovered, then merging will occur. When the sum of all token is positive this indicates that many free places are taken. If too many positions are taken, a new peer must be inserted. On the other hand, the sum of all token is minus, this indicates that too many free places exist. If too many free places, a peer shall be removed.

Three decentralized operations which may occur on any peer:

Algorithm 1. Entries inserted algorithm

1:	P ← the peer which is used for entries storage	33:	if $E \geq E_{max}$ then
2:	C ← the usable memory capacity	34:	if P = the last peer of a ring-list then
3:	M ← the maximal memory capacity	35:	insert E to this peer
4:	E_{min} ← the first entry of sorting results on each peer	36:	T_i ← 1
5:	E_{max} ← the last entry of sorting results on each peer	37:	if C = M then
6:	E ← a new coming entry	38:	move E_{min} to its neighbour on the left
7:	T_i ← a token	39:	end then
8:	**Cycle**	40:	sorting all entries of this peer
9:	**wait for** the coming entry	41:	update E_{min} and E_{max}
10:	check emergency rule	42:	end then else
11:	if $E < E_{min}$ then	43:	send E to its neighbor on the right
12:	if P = the first peer of a ring-list then	44:	end else
13:	insert E to this peer	45:	end then
14:	T_i ← 1	46:	if E_{max} come then
15:	if C = M then	47:	if C < M then
16:	move E_{max} to its neighbour to the right	48:	insert E_{max} to this peer
17:	end then	49:	sorting all entries of this peer
18:	sorting all entries of this peer	50:	update E_{min} and E_{max}
19:	update E_{min} and E_{max}	51:	end then else
20:	end then else	52:	send E_{max} to its neighbor on the right
21:	send E to its neighbour on the left	53:	end else
22:	end else	54:	end then
23:	end then	55:	if E_{min} come then
24:	if $E \geq E_{min}$ and $E < E_{max}$ then	56:	if C < M then
25:	insert E to this peer	57:	insert E_{min} to this peer
26:	T_i ← 1	58:	sorting all entries of this peer
27:	if C = M then	59:	update E_{min} and E_{max}
28:	move E_{min} to its neighbour left	60:	end then else
29:	end then	61:	send E_{min} to its neighbour on the left
30:	sorting all entries of this peer	62:	end else
31:	update E_{min} and E_{max}	63:	end then
32:	end then	64:	end cycle

Algorithm 2. Tokens combination algorithm

1:	T_i , T_j ← an integer value of token i, j
2:	INS_p ← a constant of the memory capacity of peer p (usually set equal to 80% of the memory capacity of the peer)
3:	DEL_p ← a negative constant of the memory capacity of deleted peer p
4:	**Cycle**
5:	set a randomly chosen time and **wait for** tokens to meet
6:	for j ← 1 to N do
7:	extend the time (according to the number of tokens)
8:	T_i ← T_i + T_j (T_j will be removed)
9:	end for
10:	if $T_i \geq INS_p$ then
11:	insert a new peer p'
12:	T_i ← T_i + $INS_{p'}$ ($INS_{p'}$ represent the space on the new peer p')
13:	end then
14:	if $T_i \leq DEL_p$ and the peer p is not the last peer then
15:	forward token without any operation and move the data items to a neighbour on the right peer
16:	if move successfully then
17:	T_i ← T_i + DEL_p
18:	move T_i to a neighbour on the right peer and delete this peer
19:	end then
20:	end then
21:	move tokens to a neighbor on the right peer
22:	end cycle

1. Insert operation is activated when a new item is stored, a token ($T_i = 1$) is generated and the items of this ring-list are in a lexical order following the existing order relation $>$, $<$, $=$ the keys of all entries, see in Algorithm 1.
2. Remove operation is operated when an entry is deleted ($T_i = -1$ is added).
3. Combine operation is activated when two or more tokens encounter each other on the same peer before being merged together ($Ti = Ti + Tj$ and token j will be removed), see in Algorithm 2 and Fig. 3.

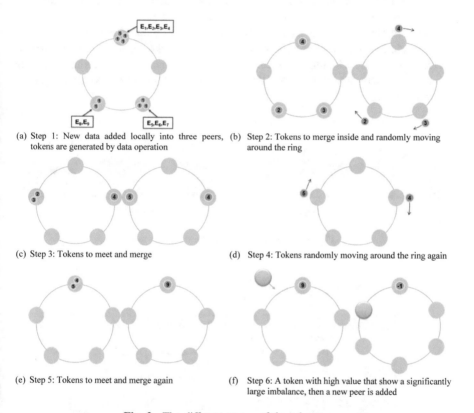

(a) Step 1: New data added locally into three peers, tokens are generated by data operation

(b) Step 2: Tokens to merge inside and randomly moving around the ring

(c) Step 3: Tokens to meet and merge

(d) Step 4: Tokens randomly moving around the ring again

(e) Step 5: Tokens to meet and merge again

(f) Step 6: A token with high value that show a significantly large imbalance, then a new peer is added

Fig. 3. The different stages of the token game

In Algorithm 2 and Fig. 3, if a token with a high value $T_i \geq INS_p$ this means that many places are taken on the peer exists, a new peer is added to a ring-list and to reduce the token value by $T_i := T_i - INS_{p'}$. Likewise, if a token value $T_i \leq DEL_p$ this means that too many free places are unoccupied, the peer p on which the token is located manages (i.e. a peer pointing on itself as predecessor and successor) then to move both tokens and the data items to its neighbour right peer. After that, this peer is removed from the ring-list and to increase the token value by $T_i := T_i + DEL_p$.

Furthermore, *the emergency rule* is activated if the capacity of any peer is used and an entry insertion is requested continuously, a new peer will be immediately added, to divide entries of overfull peer into two equal halves and move to a new peer, and then to set a value $= -INS_{p'}$, where $INS_{p'}$ represent the space on the new peer p' after that, this value with T_i by $T_i := T_i - INS_{p'}$. will be added later.

2.3   Load Balancing Mechanism

A ring structure contains the number of peers is running independently from each other. Each peer communicates with its direct neighbours and an initial load of the list of entries to be performed. Then, each peer may have a different processing capacity, moreover, remaining usable capacity of each peer will be different as well. As a result, the processing time is interrupted and the speed to execute each task has been minimized.

Hence, to reach good access speed and performance, the load distribution strategies and techniques should be implemented. The conceptual idea of this research is based on the module of communication tubes; the liquid will reach the same level in all parts of the structure, see Fig. 4.

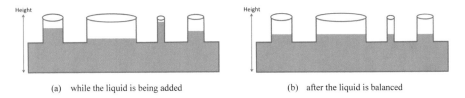

(a)   while the liquid is being added   (b)   after the liquid is balanced

Fig. 4. The filling of communicating tube

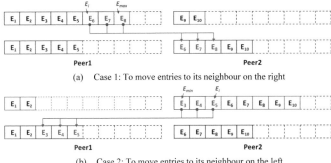

(a)   Case 1: To move entries to its neighbour on the right

(b)   Case 2: To move entries to its neighbour on the left

Fig. 5. Example: two case of balancing the number of entries on a ring-list

Different loads affect the communication of each peer on the ring-list with its neighbours. If a peer contains more entries of β percent than its neighbours, entries will be sent to one neighbour peer with the highest difference of items in order to equalise the loads (Note that if the selected peer to send entries is its neighbour right, entries will begin to move from E_i to E_{max} (see in Fig. 5(a)), but if it is its neighbour left, E_{min} to E_i will be moved (see in Fig. 5(b))). By doing so, it must be ensured that the right lexical order is kept.

Time	Peer1	Peer2	Peer3	Peer4	Peer5
0	20	**80**	0	20	20
1	20	**40**	**40**	20	20
2	30	30	30	**30**	20
3	30	30	**30**	25	25
4	30	**30**	27	**28**	25
5	**30**	28	**29**	26	27
6	29	**29**	27	28	27
7	29	28	28	28	27

Fig. 6. Example: balancing the number of entries on a ring-list with 5 peers

In Fig. 6, After a ring is stable, Peer 2 appears to have significantly large imbalance powers, then the load balance algorithm as shown in Algorithm 3 is activated. This algorithm will be run until the powers of the existing peers on the ring-list are balanced.

Algorithm 3. Load balance algorithm

```
 1:   P ← the peer which is overload
 2:   L ← the current load of a peer
 3:   L_left ← the current load of neighbour on the left
 4:   L_right ← the current load of neighbour on the right
 5:   L_move ← the number of entries will be moved
 6:   Cycle
 7:     Wait for load change
 8:       If load change then
 9:         L_dl ← L - L_left
10:         L_dr ← L - L_right
11:         if L_dl > 1 and L_dl > L_dr then
12:           if P = the first peer of a ring-list then
13:             L_move ← L - average(L + L_right)
14:             move difference loads (L_move) to neighbour on the right
15:           end then else
16:             L_move ← L - average(L + L_left)
17:             move difference loads (L_move) to neighbour on the left
18:           end else
19:         end then
20:         if L_dr > 1 and L_dr ≥ L_dl then
21:           if P = the last peer of a ring-list then
22:             L_move ← L - average(L + L_left)
23:             move difference loads (L_move) to neighbour on the left
24:           end then else
25:             L_move ← L - average(L + L_right)
26:             move difference loads (L_move) to neighbour on the right
27:           end else
28:         end then
29:       end then
30:   end cycle
```

Now, all algorithms have been implemented in the simulation, the results will be described in the next sections.

3 Experimental

3.1 Goals

The presented algorithms will be evaluated in the experiments for a new ring-like associative memory. To adapt the ring size and balance loads, the goals of the algorithm are:

1. A new peer may participate if too many positions have been taken on the existing peer.
2. The system behaved dynamically even if new elements are inserted or deleted which also be depending on its size.
3. A peer will be left without attracting trouble if too many free places i.e. never an item will be lost during moving from a departed peer to its neighbour peer.
4. The load balancing algorithm can react a large imbalance is accumulated at one point.

To prove this as indicated goals above, the experiments will be described below.

3.2   Experimental Setup

The experiment has been performed on the simulation to show that the goals of the algorithms have been achieved. A ring structure start with only one peer at the beginning, then a list of randomised entries is inserted into a ring-list. The system flow processes shows in Fig. 7.

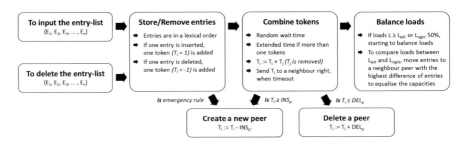

Fig. 7. The system flow processes

In Fig. 7, the number of entry-list is generated randomly. Each insertion of entry-list is processed as follows:

1. *Store entries:* an entry-list, forming a linear, is added into a ring-list. Each entry must be compared and inserted or moved to the correct positional place (a lexical order). When one entry is inserted, one token ($T_i = 1$) is added.
2. *Combine tokens:* each peer has a random time to wait for tokens to meet and this time will be extended. If more than one tokens in the same peer are found, the tokens will be merged ($T_i = T_i + T_j$, token j will be removed). After that, if this time is running out, then the token T_i is forwarded to a neighbour on the right.
3. *Create a new peer:* if the capacity of any peer is used and an entry insertion is requested, two cases of adding a new peer, the emergency rule and a value of the token $T_i \geq INS_p$, then balancing the load of a peer and a new peer. After that, this value with T_i by $T_i := T_i - INS_{p'}$. will be added.
4. *Balance loads:* after the ring is stable, load balancing algorithm will be activated if some peer pertains a significantly large imbalance load ($L \geq L_{left}$ or L_{right} 50%).

Loads of each peer is compared with its neighbours and to select the highest difference of items to equalise the powers.

In the same way, if some items are removed from the ring-list. When one entry is removed, one token ($T_i = -1$) is inserted then to combine tokens and extend times if there are more than one token in the same peer. Consequently, a peer is to be deleted if the token $T_i \leq DEL_p$. After that, this value with T_i by $T_i := T_i + DEL_p$. will be added. Finally, to balance loads if some peer has a significantly strong power as well.

3.3 Results and Discussion

Results of experiment discussed herein revealed that the algorithms which have been applied to manage the ring-list were activated i.e. when entries were inserted, tokens were met on the same peer, and load of the existing peers on the ring lost its balanced state. To ensure that the algorithms perform well as the data distributions change, the entry-list was set and deleted continuously. The results were presented in Figs. 8 and 9. After the entry-list were consistently added to the ring-list, each entry was stored in the right lexical order and the size of the ring increased gradually. When too many positions had been taken, a new peer was added to prevent excessive free places on the peer. Furthermore, the load balancing algorithm reacted to balance loads when some peer had a large imbalance load, as shown in Figs. 10 and 11.

1. Exp.1: Processing Time of Tokens

At the beginning, a ring structure start with one peer, the orange line showed that, after the first entry-list was added, (Time = 1), tokens were inserted (one item generate one token and the value of token $T_i = 1$). Therefore, the results of the number of tokens decreased sharply during this first time. Consequently, when tokens were rotating around the ring, they will be met on some peer and then will be merged together ($T_i = T_i + T_j$, token j will be removed). Thus the number of tokens dropped slightly and remained constant at later times. When the next entry-list were initiated (i.e. Time = 15, 31, 50, 93, 151, 196, 244, 295, and 342) the tokens were generated, met, and merged again. Moreover, during insertion of the entry-list, if some value of tokens was presented in high value, then a new peer was added in order to maintain the results of the number of peers as increasing as indicated by the blue line, see in Fig. 8 for more details.

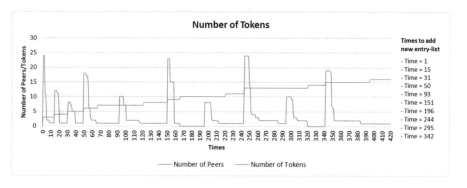

Fig. 8. Processing time of the number of tokens with 10 entry-lists insertion

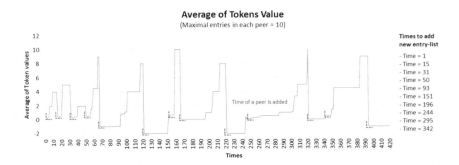

Fig. 9. Average of tokens value with 10 entry-lists insertion

Figure 9 presents, in insertion entry-lists, that each peer can contain a maximum of 10 items. When each entry-list was added, tokens were inserted. While entry-lists were being inserted, if the value of some of the token were high ($T_i \geq INS_p$), then a new peer is added, which is illustrated by dashed red line. The tendency of all entry-lists insertion were the same, after many tokens were merged then the average value of all tokens dramatically increased. When some tokens contain a high value and new peers were inserted, the average of the tokens value was significantly dropped. The average value of all tokens was equal to or less than 10 and above −2 which means that no excessive places exist on the peer.

2. Exp.2: Processing Time of Load Balancing

In each insertion and deletion of the entry-list, when items are inserted into the peers, some of these peers may be larger than the others. Consequently, the load balancing algorithm will works when the number of the entries contains a large imbalance load, see in Figs. 10 and 11.

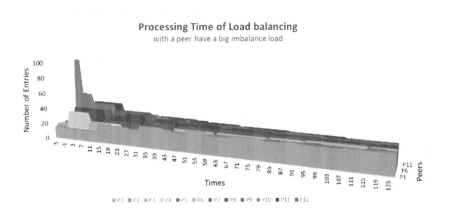

Fig. 10. Processing time of load balancing with a peer which had large imbalance load

The graph in Fig. 10 represents a stabilized ring, entry-lists are randomly inserted to the ring, see in Table 1 (Time = 0).

Table 1. Number of entries at Time = 0 and Time = 125

Peer	P1	P2	P3	P4	P5	P6	P7	P8	P9	P10	P11	P12
#Entries (Time = 0)	14	13	13	35	17	17	7	37	36	100	10	13
#Entries (Time = 125)	26	25	25	24	24	26	26	27	27	28	27	27

Then, the load balancing algorithm was activated because the number of entries of some peer, i.e. Peer 10 shows a large imbalanced load. To initiate the balance load in each peer, compare the loads with its neighbours and to select the highest difference of items, to send the entries for equalising the powers, and repeated until the number of entries of the existing peers on the ring-list is balanced, the results of this are shown in Table 1 (Time = 125).

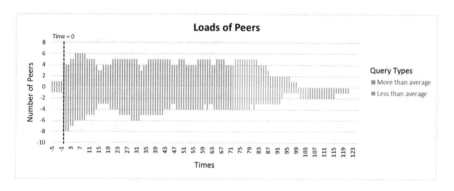

Fig. 11. Loads of the peers

Figure 11 shows the number of peers which can be more or less than the average of the loads. According to the graph, after the load balance algorithm was activated (Time = 1), the number of peers which more or less than the average declined gradually until approaching zero which means that the load is reaching the closing balance.

4   Conclusion

The algorithms and implementation details of the dynamic data management for an associative P2P memory have been presented. These developed algorithms worked on the different sizes of the entry-list which were inserted or deleted into a ring-list. All items were assigned by no fixed peer. The tokens were created and forwarded around the ring to update the state of the system. The ring was growing or decreasing without having too many free places on the peer. Memory cells are added when required by the system. Decentralised token game was shown as being capable of determining the right moment for doing so and these load balancers ensure the equal load distribution. Furthermore, these load balancers may also react to user access imbalance. Overall, the search function could be accomplished at a faster rate.

References

1. Lv, Q., Cao, P., Cohen, E., Li, K., Shenker, S.: Search and replication in unstructured P2P networks. In: Proceedings of the 16th International Conference on Supercomputing, pp. 84–95. ACM, New York (2002)
2. Jin, X., Chan, S.-H.G.: Unstructured peer-to-peer network architectures. In: Handbook of Peer-to-Peer Networking, pp. 117–142. Springer, Boston (2009)
3. Castro, M., Costa, M., Rowstron, A.: Peer-to-peer overlays: structured, unstructured, or both. Technical report MSR-TR-2004-73, Microsoft Research, System and Networking Group, Cambridge, UK (2004)
4. Stoica, I., Morris, R., Karger, D., Kaashoek, M.F., Balakrishnan, H.: Chord: a scalable peer-to-peer lookup service for internet applications. In: Proceedings of the 2001 Conference on Applications, Technologies, Architectures, and Protocols for Computer Communications, pp. 149–160. ACM, New York (2001)
5. Karger, D., Lehman, E., Leighton, F., Levine, M., Lewin, D., Panigrahy, R.: Consistent hashing and random trees: distributed caching protocols for relieving hot spots on the World Wide Web. In: Proceedings of the 29th Annual ACM Symposium on Theory of Computing, pp. 654–663. ACM, New York (1997)
6. US Secure Hash Algorithm 1 (SHA1). http://www.ietf.org/rfc/rfc3174.txt. Accessed 20 Mar 2018
7. Siamak, S.: A peer-to-peer dictionary using chord DHT. http://citeseerx.ist.psu.edu/viewdoc/versions?doi=10.1.1.173.7388. Accessed 20 Mar 2018
8. Pitkow, J.E.: Summary of WWW characterizations. In: 7th International World-Wide Web Conference, Brisbane, Australia, pp. 3–13 (1998)

Extremely Fast Neural Computation Using Tally Numeral Arithmetic

Kosuke Imamura[✉]

Eastern Washington University, Cheney, WA 99004, USA
kimamura@ewu.edu

Abstract. The tally numeral is an ancient and simple counting system. Yet, it achieves extremely fast neural computation by combinatorial logic. This paper introduces the tally numeral arithmetic logic and shows that the tally numeral AND operation performs as a Rectified Linear Unit, suggesting its potential applications to deep neural networks. The first tally numeral arithmetic ADD and AND were implemented on a Field Programmable Gate Array (FPGA) in 2016 to build a perceptron. We prototyped a three-layer XOR network to exercise a fast and flexible tally numeral multiplier, which can perform non-linear transformation of input values as well.

Keywords: Hardware neural networks · FPGA · Tally numerals

1 Introduction

Perceptrons (or artificial neurons) typically perform two computations: the summation of weighted inputs and the output calculation. Either for hardware or software, these computations are a bottle neck for faster execution. This paper proposes a tally numeral arithmetic implemented by combinatorial gate logic to speed up these computations to several gate delays.

The first tally numeral based perceptron was prototyped in 2016 on an Altera Cyclone III Field Programmable Gate Array (FPGA) [1]. The prototype named "CL-Perceptron" completes the de-facto standard three layer XOR neural computation in less time than two integer additions. Table 1 gives the estimated gate count and execution speed. The gate count of ATOM is listed as a reference to show that roughly 12,500 two-input XORs can be populated in the ATOM processor's die space.

This paper is organized as follows: Sect. 2 reviews hardware based neural networks and shows the relation between a Rectified Linear Unit (ReLU) and the tally numeral weight-ANDing scheme. Section 3 introduces the data representation and arithmetic operations. Section 4 describes an experimental three layer XOR network with the tally numeral multipliers. Section 5 discusses characteristics, advantage and disadvantage of the tally numerals, and our plan for the future development.

© Springer Nature Switzerland AG 2020
P. Boonyopakorn et al. (Eds.): IC2IT 2019, AISC 936, pp. 205–212, 2020.
https://doi.org/10.1007/978-3-030-19861-9_20

Table 1. The execution speed and gate count of CL-Perceptron, taken from [1].

	Estimated gate count	Execution speed
CL-Perceptron XOR	960	14.17 ns
Simple integer add	108	8.13 ns
ATOM(2008)	12,000,000	–

2   Related Works

2.1   Hardware Neural Networks

There are many types of hardware neural networks. Skrbek implemented all needed functions at a gate-level by adders and barrel shifters with "Shift-Add" arithmetic for faster computation [2]. Parthasarathy et al. replaced the real number multiplication with shifting operation by making connection weights powers of two to increase speed [3]. For hardware implementation of neural networks to speed up the computation, Field Programmable Gate Arrays (FPGA) are often used [4–7].

2.2   Rectified Linear Unit

This section shows that the weight-ANDing used in the CL-Perceptron behaves as the Rectified Linear Unit (ReLU). The applications of ReLU are found in deep neural networks [8–12]. The tally numeral weight-ANDing performs ANDing of an input and the associated weight instead of conventional multiplication or input weighting.

Although the weight-ANDing scheme was developed independently of ReLU to speed up the computation, the weight-ANDing is equivalent to ReLU. In the tally numeral, the numbers are represented by tallying the number of bits of value 1. For example, in a 6-bit system, positive 4 is 001111 (next section explains the arithmetic). The weight-ANDing is simple bitwise ANDing of input and the corresponding weight, i.e.,

$$f(x) = min(weight, x),$$

which can be transformed to ReLU by translation and reflection, i.e.,

$$f(x) = -1*min(0, weight - x) + bias.$$

Since the weight-ANDing caps the output by the weight, its behavior becomes equivalent to ReLU Capped-6 [12] with appropriate translation of bias (See Fig. 1). This suggests that the tally numeral ANDing scheme may find its use in deep neural networks where ReLU is used. The tally numbers and weights can be either positive or negative. Therefore, the weight-ANDing outputs a positive or negative value.

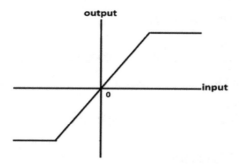

Fig. 1. Graph of the CL-Perceptron's weight-ANDing.

3 Data Representation and Arithmetic

This section introduces the data representation and arithmetic of the tally numeral system.

3.1 Data Representation

The tally numeral represents a number by tallying the number of bits of value 1. A tally number is represented by a vector of contiguous n bits. The vector$[n-1]$ represents the sign, 0 being positive and 1 being negative. A positive number counts 1 bit from the vector$[0]$ bit toward the vector$[n-1]$, while a negative number counts 1 bit from the vector$[n-1]$ toward the vector$[0]$. The numbers range from $-n$ to $n-1$.

For example, in a 9-bit system, 000001111 is +4 and 111100000 is −4, the vector $[0]$ being the rightmost. The numbers ranges from −9 to +8.

3.2 Arithmetic

A barrel shifter and OR operation implement an adder and bitwise ANDing implements weight multiplication. There are 3 types of add operations for different sign combinations: (1) two numbers are both positive, (2) one number is positive and the other is negative, and (3) two numbers are both negative. Examples of 9-bit tally numeral arithmetic are shown below:

Two Positive Numbers, A and B:

sum = (A << B) | B;
Example: 011111111 = 000111111 + 000000011

Positive number A and negative number B:

if(abs(A) > abs(B)) sum = A >> B;
else sum = B << A;
Example: 111100000 = 000000011 + 111111000

Two Negative Numbers A and B:

sum = (A << B) | B;
Example: 111111110 = 111111000 + 110000000

Weight multiplication (weight-ANDing):

Product = A & B
Example: 000000011 = 000000011 & 000111111

3.3   Tally Numeral Multiplier

The tally numeral multiplier uses a mapping function to perform multiplication of two absolute numbers. The sign of the product is computed by XORing the sign bits. Mapping is implemented by ANDing inputs with the 0/1 matrix entries and ORing the ANDed results as shown in Fig. 2. For example, the *output[1]* is calculated by ANDing each input bit with the corresponding bit in the 2nd row from the bottom of the 0/1 matrix, then ORing all the ANDed values. This mapping operation is expressed as:

$$output[i] = OR_{j=0 \ to \ N-1}(input[j] \ \& \ matrix[i][j]),$$

where *input[]* and *output[]* are input/output bit vectors respectively. The *matrix[0][0]* is at the bottom-left in the 0/1 matrix in Fig. 2. This 0/1 matrix can be used to implement S-shaped ReLU and piecewise linear activation functions proposed in [11].

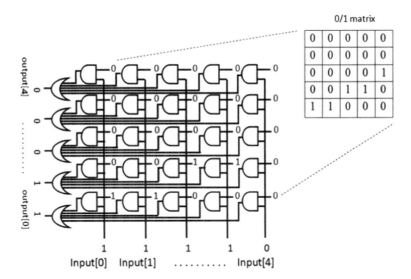

Fig. 2. The tally numeral multiplier and an example of the 5 × 5 0/1 matrix.

4 Experiment of the Multiplier with CL-Perceptron

Our previous CL-perceptron prototype is equipped with a linear-feedback shift register for the online training. However, in this experiment, we used the tally numeral multiplier instead of weight-ANDing, and offline training. The entire system was tested by software simulation.

Our test network is a de facto standard three layer XOR network shown in Fig. 3. The input and output are 16-bit tally numeral numbers. The 0/1 matrices (multiplier functions) were created by a random search. The logical 0 and 1 are coded −1 and 1 respectively. Note that the weight set, W = {w0, w1, w2, w3, w4, w5}, is not a set of constants. It is a set of increasing functions (see Fig. 4). The functions are generated randomly roughly along the following linear equations:

$$y = ax, \text{ where } a = 1/16, 3/16, 5/16, \text{ and } 7/16.$$

The network was trained with a data set {(0, 0), (0, 1), (1, 0), (1, 1)}, where the input value 0 and 1 are coded 0000000000000000 (=0) and 0011111111111111 (=14) respectively. The training generated a set of the 6 weight multiplier functions shown in Fig. 4. The functions, w2, w3, and w5, will act the same as multiplication by a constant weight. This XOR attained generalization treating the input values from 0 to 6 as 0 and from 13 to 16 as 1. Generalization is ability to classify data not included in the training set. The execution speed of the multiplier is 3 three-input OR gates plus 1 AND gate delays.

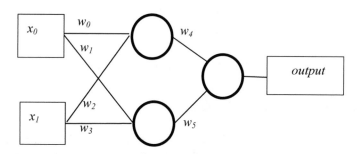

Fig. 3. The XOR network built with the tally numeral multiplier functions.

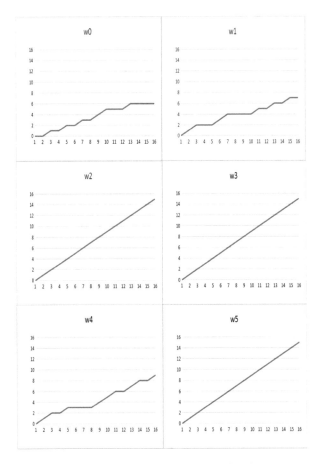

Fig. 4. The tally numeral multiplier functions.

5 Remarks and Future Development

We built a de facto standard XOR network to exercise the tally numeral arithmetic. Addition is implemented by a barrel shifter, and multiplication is implemented by the AND/OR mapping function. Conventional neural networks perform linear transformation of inputs by multiplying constant weights. The tally numeral multiplier maps input to output via a linear or non-linear increasing function.

The speed of multiplication is within several gate delays. For n-bit tally number multiplication, the total gate delay of the tally numeral multiplier is one AND gate delay plus $log_3 n$ OR gate delays, assuming that a 3-input OR gate is used. For example, an 81-bit tally numeral multiplication requires only 5 gate delays. For comparison, a 16 bit adder using four 4-bit carry lookahead units requires 8 gate delays. The tally numeral addition is as fast as the barrel shifter.

The tally numerals are data error resilient. In the binary system, the value would be considerably different if a bit error occurs at a higher-order bit position. However, the difference will be only 1 for the tally numerals no matter where a bit error occurs, and the error will be repaired except when an error occurs at the 0/1 border. For instance, 0011011 can be restored to 0011111, but a bit error can not be detected for the numbers 0011111 and 0001111. Yet, the difference of these two numbers is only 1 while it would be 16 for a binary number.

Disadvantage of the tally numeral arithmetic is that the die space does not scale. The die space growth of the 0/1 matrix is $O(n^2)$ for n-bit numbers. However, since the 0/1 matrix is a sparse matrix, we could use a two-tuple list of row and column indexes and the multiplier could be divided into multiple smaller blocks. Although the coarse resolution is thought to be disadvantage, we speculate appropriate quantization may contribute to better generalization. Further studies are needed on this issue.

Current multiplier functions are limited to be non-decreasing. This limitation can be removed by capping the output by the highest index i, such that $input[i] = 1$. We speculate non-linear input transform reduces the number of neurons that a conventional neural network requires, since conventional neural networks combine linear transforms to implement non-linear transform. Again, further analysis and experiment is needed. The tally numeral AND operation functions as a Rectified Linear Unit.

We are currently porting a tally numeral based neural network onto Altera DE10 SoC/FPGA to perform handwriting recognition as a practical application. We will be using CPU to train the network and FPGA to carry out the neural computation.

References

1. Imamura, K.: Combinatorial digital logic perceptron. In: Proceedings of 2nd International Conference on Information Technology and Computer Science, Pattaya, Thailand, pp. 55–59 (2016)
2. Skrbek, M.: Fast neural network implementation. http://www.academia.edu/4397771/FAST_NEURAL_NETWORK_IMPLEMENTATION. Accessed 22 Dec 2018
3. Parthasarathy, V., Hemapriya, B.C., Ravikumar, H.M.: A simplified design of multiplier for multi-layer feed forward hardware neural networks. Int. J. Res. Eng. Technol. **3**, 43–47 (2014). Special Issue 12, ICAESA. http://www.academia.edu/8034384/A_SIMPLIFIED_DESIGN_OF_MULTIPLIER_FOR_MULTI_LAYER_FEED_FORWARD_HARDWARE_NEURAL_NETWORKS. Accessed 22 Dec 2018
4. Ortega-Zamorano, F., Jerez, J., Muñoz, D.U., Luque-Baena, R.M., Franco, L.: Efficient implementation of the backpropagation algorithm in FPGAs and microcontrollers. IEEE Trans. Neural Netw. Learn. Syst. **27**(9), 1840–1850 (2016). http://ieeexplore.ieee.org/xpl/articleDetails.jsp?arnumber=7192642. Accessed 22 Dec 2018
5. Jung, S., Kim, S.: Hardware implementation of a real-time neural network controller with a DSP and an FPGA for nonlinear systems. IEEE Trans. Ind. Electron. **54**(1) (2007) http://ieeexplore.ieee.org/xpl/abstractAuthors.jsp?arnumber=4084734. Accessed 22 Dec 2018
6. Deotale, P.D., Dole, L.: Design of FPGA based general purpose neural network. In: International Conference on Information Communication and Embedded Systems, pp. 1–5 (2014). https://ieeexplore.ieee.org/stamp/stamp.jsp?tp=&arnumber=7033843. Accessed 1 Feb 2019

7. Pietras, M.: Hardware conversion of neural networks simulation models for neural processing accelerator implemented as FPGA-based SoC. In: 24th International Conference on Field Programmable Logic and Applications (2014). http://ieeexplore.ieee.org/xpl/articleDetails.jsp?arnumber=6927383&queryText=FPGA%20Neural%20Networks&newsearch=true. Accessed 1 Feb 2019

8. Glorot, X., Bordes, A., Bengio, Y.: Deep sparse rectifier neural networks. http://www.utc.fr/∼bordesan/dokuwiki/_media/en/glorot10nipsworkshop.pdf. Accessed 22 Dec 2018

9. Agarap, A.F..: Deep learning using rectified linear units (ReLU) (2018). https://arxiv.org/pdf/1803.08375.pdf. Accessed 30 Jan 2019

10. Xu, B., Wang, N., Chen, T., Li, M.: Empirical evaluation of rectified activations in convolutional network (2015). https://arxiv.org/pdf/1505.00853.pdf. Accessed 31 Jan 2019

11. Jin, X., Xu, C., Feng, J., Wei, Y., Xiong, J., Yan, S.: Deep learning with S-shaped rectified linear activation units. In: Proceedings of the Thirtieth AAAI Conference on Artificial Intelligence, Phoenix, AZ USA, pp. 1737–1743 (2016). https://www.aaai.org/ocs/index.php/AAAI/AAAI16/paper/download/12358/11798. Accessed 31 Jan 2019

12. Krizhevsky, A.: Convolutional deep belief networks on CIFAR-10 (2016). https://www.cs.toronto.edu/∼kriz/conv-cifar10-aug2010.pdf. Accessed 31 Jan 2019

Traceable CP-ABE for Outsourced Big Data in Cloud Storage

Praveen Kumar Premkamal[1,2]([✉]), Syam Kumar Pasupuleti[2],
and P. J. A. Alphonse[1]

[1] National Institute of Technology, Tiruchirappalli, India
tvp.praveen@gmail.com, alphonse@nitt.edu
[2] Institute for Development and Research in Banking Technology, Hyderabad, India
psyamkumar@idrbt.ac.in

Abstract. Ciphertext Policy Attribute Based Encryption (CP-ABE) is
a promising cryptographic solution for the unauthorized access to the
outsourced big data in the cloud environment. However, most of the
existing CP-ABE schemes do not have the provision to trace the users
who misuse their secret key for profit intention which indeed reduces the
CP-ABE schemes security. Thus, in this paper, we propose traceable CP-
ABE (T-CP-ABE) for outsourced big data in cloud storage. Our scheme
has the provision to dynamically trace who is decrypting the ciphertext
during outsourced proxy decryption, which helps to identify the malicious
users who misuse their secret key for profit intention. Furthermore, our
scheme has an efficient key sanity check, which ensures that only the well-
formed secret key is used for decryption. Security analysis proves that
our scheme resist against the secret key forging and chosen-plaintext
attacks. Performance evaluation proves that our T-CP-ABE scheme is
efficient than other traceability schemes.

Keywords: Privacy · Access control · CP-ABE · Traceablity ·
Cloud storage · Big data

1 Introduction

Cloud storage provides massive and scalable storage resources as a service to
meet the on-demand needs of the organization, and it also has the benefits of
anywhere and anytime access, availability, reliability and collaboration. Since
the cloud storage offers various services in pay-per-use model, it is the best and
cost effective solution for the organizations to outsource their big data, since big
data requires huge resources to store and maintain on the premises. However,
when outsourcing the big data into the cloud there are two important security
considerations to be addressed such as the data privacy and access control [1,2].

The traditional solution for privacy in the cloud is to store the encrypted
data, but it does not provide access control which is essential for the outsourced
big data. The CP-ABE provides the attributes based effective fine-grained access

© Springer Nature Switzerland AG 2020
P. Boonyopakorn et al. (Eds.): IC2IT 2019, AISC 936, pp. 213–226, 2020.
https://doi.org/10.1007/978-3-030-19861-9_21

control in the cloud environment. In CP-ABE, the data owners outsource their data in an encrypted form to the cloud storage and only shares it with the authorized data users whose attributes satisfies the access policy. Here, the user secret key or decryption key and access policy consists of attributes. Data owners use the access policy to encrypt the data and data users use secret key to decrypt the ciphertext.

CP-ABE is a reliable technique to protect the outsourced big data in cloud storage. However, there is a significant security issue in CP-ABE called secret key leakage or privilege abuse that has to be addressed in an effective way. In CP-ABE, more than one user may have the same set of attributes which implies that many users can have the same secret key. Since, many users have the same secret key, it is difficult to trace whose secret key is used to decrypt the ciphertext because the secret key is associated only with attributes not with the user identity. For example, the data is encrypted with access policy (Cloud AND (Researcher OR Faculty)) by the data owner and uploaded it into the cloud. Kanishk and Priya are two users are having the same attributes (Cloud, Faculty) which satisfies the access policy. In this scenario, it is not possible to trace whether Kanishk decrypted the data or Priya decrypted the data because both the user have the same secret key. Due to this fact, the malicious users can intentionally sell their secret key to others for profit which leads to the security breach in CP-ABE schemes. This process is called secret key leak or privilege abuse. Hence, there should be a mechanism to find out whose secret key is used for decryption in order to overcome the above said security issue. This process is known as traceability.

To address this secret key leakage issue, Liu et al. [3] proposed solution to trace the user identity based on identity table which is used to map the user details with user specific information during the tracing. But the concept of identity table is not efficient because identity table grows with respect to the number of users which leads to additional overhead. Ning et al. [4,5] enhanced the earlier scheme and they introduced the tracing based on Shamir secret sharing value which is also computationally inefficient. To improve the efficiency, Zhang et al. [6] added the user identification as one parameter in the user key, and they found it during tracing, however this approach is not secure because malicious user may modify the parameter which makes it hard for data owner to identify the exact malicious users. Recently Ning et al. [7–9] proposed improved versions of traceability in which they embedded the user identity with user secret key, and during the tracing they extract the user identity. However these schemes are not suitable for practical cloud applications because of the following reasons.

1. Extracting the user identity from secret key is carried out after the decryption process which brings several issues that was not addressed such as (i) How the abnormal data access found? (ii) Who maintained the secret keys that were used during decryption? (iii) Where were the secret keys stored? (iv) What about the storage and communication overhead if tracing was performed by other than data user?

2. The key sanity check process suffered with computation overhead because it involves more number of pairing operations.

Hence, it is required to design dynamic practical traceable system for CP-ABE schemes. Thus, we propose traceable CP-ABE for outsourced big data in cloud storage. Our contribution is as follows:

1. *Dynamic traceability.* We propose a dynamic traceability scheme in which the proxy server extracts the user identity from secret key during the outsourced decryption process. Since the tracing process is carried out during the outsourced decryption itself by PS, the secret key does not have to be stored and communicated to any other entity which indeed reduces the storage overhead and the communication overhead.
2. *Identifying the abnormal date access.* After extracting the user identity, PS stores the user identity in the log file. The log file is available with the proxy server which helps to identify the malicious user who accessed the data abnormally.
3. *No additional storage needed for tracing.* We embed the user identity in the user secret key itself. During the tracing process it is extracted directly from the secret key, thus there is no need for maintaining the identity table to get the user identity as like in the scheme [3].
4. *Efficient key sanity check.* We improve the efficiency of the key sanity check process by reducing the pairing operations required.
5. *Reduce the data user computation overhead.* Decryption process requires more computation overhead in CP-ABE schemes that is performed by the data users. To reduce the data user computation overhead, we outsourced all inflated pairing computations to the proxy server.

Table 1 gives the comprehensive comparison of different traceability CP-ABE schemes. It is clearly observed from Table 1 that, our scheme is the only scheme which achieves dynamic traceabiity, identifying the abnormal access and efficient key sanity check.

Table 1. Comparison of CP-ABE with traceability schemes

Schemes	TR	NSTR	DYNTR	ADA	EKC
LCW [3]	✓	×	×	×	×
NDCWL [4]	✓	×	×	×	×
LZYZ [10]	✓	×	×	×	×
NCDW [7]	✓	✓	×	×	×
T-CP-ABE (our scheme)	✓	✓	✓	✓	✓

TR: Traceability; NSTR - No storage for tracing; DYNTR - Dynamic Traceability; ADA - Identifying the abnormal date access; EKC - Efficient key sanity check.

2   Related Works

Bethencourt et al. [11] initially put forth the idea of CP-ABE. In their scheme, they used access tree as access policy. After that, AND gate as the access structure was proposed in [12], CP-ABE with matrix based access structure Linear Secret Sharing Scheme (LSSS) was proposed in [13]. Recently, another new access structure Ordered Binary Decision Diagram (OBDD) was proposed in [14]. To improve the efficiency, CP-ABE towards short ciphertext and fast decryption was presented in [15–17] and outsourced CP-ABE schemes was proposed in [18,19]. Later, various efforts have been taken to improve the CP-ABE schemes [20–26]. However these schemes did not address the traceability.

Li et al. [27] introduced the accountable CP-ABE using AND-gate access structure to prevent the key selling or sharing misbehaviour activities among the malicious users. In their scheme, they added the extra user specific value in the secret key and it is traced during tracing process. Li et al. [28] proposed multiple-authority CP-ABE with accountability in which they have the provision to trace the misbehaved users. Both the above schemes [27,28] are based on AND-gate with wild-card access structure which is not expressive.

Later, Katz et al. [29] proposed the traceability for predicate encryption in which the group manager was able to find out whose key was used to decrypt the ciphertext. Liu et al. [3] proposed CP-ABE with traceability feature for any monotone access structure in which they added user specific information in the secret key and also they used the identity table which is used to map the user details with user specific information during the tracing. The complexity of the tracing process linearly grows according to the number of users which leads to the additional overhead. Ning et al. [4,5] overcame the complexity produced by identity table and introduced the tracing based on Shamir secret sharing value for each users which was also computationally inefficient. Ning et al. [7] improved their previous scheme and proposed a better CP-ABE scheme with traceability without using identity table and Shamir secret value. Later, Ning et al. [8,9] constructed their schemes by adding the additional auditing features to verify whether the user is leaked or not. Recently, traceable CP-ABE scheme in health care proposed by Li et al. [10]. Zhang et al. [6] proposed the white-box traceability in multi-authority CP-ABE system. In their schemes, they added the user identification as one parameter in the user key which is not secure because malicious user may modify and access the data. However all the above traceability schemes [3–10, 27–29] trace the user identity in static way which happens only after decryption process.

3   Preliminaries

Our proposed T-CP-ABE scheme designed using the basic concept of tree access structure, bilinear pairing and Shamir secret sharing and the security is proved based on Decisional Bilinear Diffie-Hellman (DBDH) assumption.

Due to space limitations, we refer the scheme [19] for detailed definitions.

4   T-CP-ABE System Model

Here, we describe the proposed T-CP-ABE scheme system model. The system model have five entities such as data owner, data users, trusted authority, proxy server and cloud server, and the Fig. 1 shows the T-CP-ABE architecture.

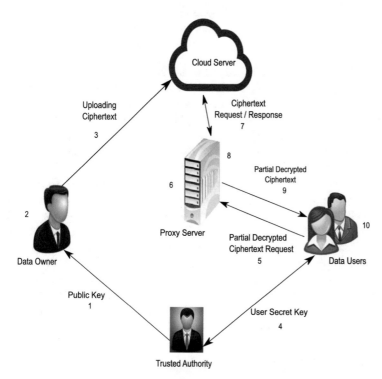

1. TA sends the public key to DO   2. DO defines access policy and encrypts the data using public key and defined access policy
3. DO uploads the ciphertext into the cloud 4. TA generates secret key and sends to users 5. DU requests the partial decrypted ciphertext from PS
6. PS checks whether the secret key is well-formed or not and also traces the user identity. 7. If the secret key is well-formed then PS reterieves the
ciphertext from the cloud. 8.PS performs partial decryption 9. PS sends the partially decrypted ciphertext to the DU 10. DU decrypts

Fig. 1. T-CP-ABE scheme system architecture

Trusted Authority (TA): The TA creates the master secret key (MSK) and public key (PK). It also creates the user secret key (US) based on master secret key for decryption.

Data Owner (DO): Trusted DO creates the access structure (τ) and the ciphertext (CT). DO outsources the encrypted data to the cloud and shares with only the authorized users based on the defined access policy.

Proxy Server (PS): The PS is a semi-trusted entity. The semi-trusted means honest, but curious which means PS performs the assigned task correctly, but it is curious to obtain the plain text. The proxy server performs the following three processes such as (i) verify whether secret key is well-formed or not (ii) trace the user who is decrypting the data (iii) partial decryption.

Cloud Server (CS): CS is an untrusted entity. The cloud has huge volume of scalable storage. It provides storage as a service to the DO and it provides the encrypted data access to the PS.

Data users (DU): The untrusted DU gets the partially decrypted ciphertext from PS and decrypts the data. Only authorized users get the exact message only when the DU attributes satisfied the access policy.

In T-CP-ABE scheme, the TA generates the MSK and PK, and sends the PK to DO for data encryption. TA retains the master key for user secret key generation. Then, the DO defines the access policy and performs data encryption using access structure and public key. The ciphertext is then uploaded into the cloud by the DO. When the DU needs the plaintext, first the user requests the secret key from TA. In response to the request, the TA creates user global secret key and user secret key based on user attributes and MSK, and it gives to the DU. Then, the DU requests the partially decrypted ciphertext from proxy server by sending the user secret key. PS first checks whether the received user secret key is well-formed or not. If it is well-formed, then the PS extracts the user identity from the secret key and stores it in the log file. The log file helps to identify the malicious user who accessed the data abnormally. After that, PS downloads the ciphertext from the cloud if the secret key is well-formed and performs partial decryption. After completion of partial decryption the PS sends the partial decrypted ciphertext to DU. DU decrypts the partially decrypted ciphertext using user global secret key. DU obtains the exact plaintext when the DU attributes satisfied the access structure.

5   T-CP-ABE Scheme Construction

Here, we construct the proposed Scheme. It consists of seven algorithms such as SystemSetup, UserSetup, Encryption, KeySanityCheck, Trace, ProxyDecryption and Decryption. The detailed construction of T-CP-ABE Scheme is given below.

5.1   SystemSetup

SystemSetup is executed by TA with security parameter (d) as input, and it creates MSK and PK. Let G be the additive cyclic group of prime order p. Let G_t be the multiplicative cyclic group of prime order p. The generator of group G is denoted by $\mathfrak{g}$, and the bilinear function ϱ is defined as $\varrho : G \times G \to G_t$. Let $\mathbb{UL}$ be the system user list, and the universal attribute set is denoted by $\mathbb{U}$. Let $H : G \to Z_p$ be the collusion resistance hash function. The detailed steps involved in creating MSK and PK are as follows:

1. Select random numbers $x_{att}, \forall att \in \mathbb{U}$ and select a random number κ from Z_p.

2. Compute X_{att} as $X_{att} = \mathfrak{g}^{x_{att}}$, where $\forall att \in \mathbb{U}$ and compute $v = \varrho(\mathfrak{g}, \mathfrak{g})^{\kappa}$
3. MSK $= ((x_1, x_2, ..., x_m), \kappa)$ and PK $= (X_{att} \colon \forall att \in \mathbb{U}, \mathfrak{g}, v, H)$
4. Return MSK, PK

After generating PK and MSK, TA sends PK to the DO and retains the MSK for user secret key creation.

5.2   UserSetup

UserSetup algorithm is executed by TA with user list $\mathbb{UL}$ as input and performs the new user registration. It generates unique identification number (uid) from Z_p for each new user, and sends the same to the concern user. It also adds the uid into the user list $\mathbb{UL}$.

5.3   KeyGen

KeyGen algorithm is executed by TA. This algorithm takes MSK, PK, uid, and user attributes (δ) as inputs and outputs user global secret key (UGSK) and user secret key (US). The procedure to generate user key is given below:

1. Pick $temp$ from Z_p and let UGSK $= temp$.
2. Pick α from Z_p.
3. Compute $US_1 = \mathfrak{g}^{\kappa * UGSK + \alpha * UGSK - uid}$ and compute $US_2 = \mathfrak{g}^{uid * H(US_1)}$
4. Compute $US_3 = \mathfrak{g}^{-\kappa * UGSK}$ and compute $US_{att} = \mathfrak{g}^{\frac{\alpha * UGSK}{x_{att}}}, \forall att \in \delta$
5. Return $US = (US_1, US_2, US_3, US_{att} \colon \forall att \in \delta)$, UGSK

At the end of key generation, the TA sends the US and $UGSK$ to the concern user.

5.4   Encryption

DO executes the Encryption algorithm for encrypting the data using PK and access structure (τ) as inputs and it outputs ciphertext (CT). Let r_o be the root node of τ and y denotes the nodes of τ. Let t_y represents the threshold value of node y. Each child node of node y is represented as cn. All the leaf nodes in τ is denoted by L. The various steps involved in the Encryption algorithm is given below:

1. Select a random number from Z_p and store it in β
2. Choose $t_{r_o} - 1$ degree random polynomial p_{r_o} and let $p_{r_o}(0) = \beta$
3. Now compute the secret share value (s_{cn}) for each child nodes of root node using Shamir secret sharing scheme. The share value $s_{cn} = p_{r0}(index(cn))$.
4. Perform the following lines from 5 to 6 repeatedly $\forall$ cn of τ.
5. if cn is a leaf node then compute $s_y(0) = p_{parents(y)}(index(y))$
6. if cn is a non-leaf node
   (a) Choose $t_y - 1$ degree random polynomial p_y

   (b) Find the share value of each cn using Shamir secret sharing scheme. Compute $s_{cn}(0) = p_y(index(cn))$.

7. $\forall y \in L$, compute $CT_{att} = (X_{att})^{s_y} = (X_{att})^{p_y(0)} = \mathfrak{g}^{p_y(0).x_{att}}$
8. Compute $CT_1 = \mathrm{M}.v^\beta = M.\varrho(\mathfrak{g}, \mathfrak{g})^{\kappa\beta}$ and $CT_2 = \mathfrak{g}^\beta$
9. Return Ciphertext $CT = (\tau, CT_1, CT_2, CT_{att}: \forall att \in L)$

After encryption, the DO uploads the CT into the cloud.

5.5   KeySanityCheck

KeySanityCheck algorithm is executed by PS. This algorithm is used to check whether the secret key is well-formed or not. It takes PK and US as input and returns the kflag as 0 or 1. If kflag $= 1$, then secret key is well-formed otherwise secret key is not well-formed. The procedure of KeySanityCheck is given below:

1. Let $A = \varrho(\mathfrak{g}, US_1 * (US_2 * \mathfrak{g}^{-H(US_1)}) * US_3) = \varrho(\mathfrak{g}, (\mathfrak{g}^{(\kappa * UGSK + \alpha * UGSK - uid)} * \mathfrak{g}^{uid}$
   $* \mathfrak{g}^{-\kappa * UGSK})) = \varrho(\mathfrak{g}, \mathfrak{g}^{\kappa * UGSK + \alpha * UGSK - uid + uid - \kappa * UGSK}) = \varrho(\mathfrak{g}, \mathfrak{g})^{\alpha * UGSK}$
2. if $\varrho(X_{att}, US_{att}) = A$, $\forall att \in \delta$ then kflag $= 1$ else kflag $= 0$
3. Return kflag

After key sanity check, the PS executes the Trace algorithm.

5.6   Trace

PS executes Trace algorithm. This algorithm extracts uid from secret key US. It takes PK, kflag and US as inputs. If $kflag = 1$ then extract uid as $uid = log_g US_2 * \mathfrak{g}^{-H(US_1)}$ and store the uid into the log file. The log file is helpful to audit if any user have accessed the data abnormally. After successfully tracing, PS downloads the CT from the cloud and executes the ProxyDecryption algorithm if the secret key is well-formed, otherwise it terminates the process.

5.7   ProxyDecryption

PS executes ProxyDecryption algorithm to perform partial decryption on behalf of DU. It takes PK, CT, and US as inputs and outputs the partially decrypted ciphertext (PDCT). The various notations used in the this algorithm are given below: Let y denotes the node of τ. Let cn denotes the set of child nodes of node y. The threshold value of node y is denoted by t_y. Let δ_y be a set of attributes of cn. Lagrange interpolation formula is used to reconstruct the share value (s_y) for all non-leaf nodes of access tree τ. The detailed step to perform partial decryption is as follows:

1. if $y \in$ leaf node
   if att(y) $\notin \delta$ then $s_y = \perp$ else $s_y = \varrho(US_{att}, CT_{att}) = \varrho(\mathfrak{g}^{\frac{\alpha * UGSK}{x_{att}}}, \mathfrak{g}^{p_y(0)x_{att}}) = \varrho(\mathfrak{g}, \mathfrak{g})^{\alpha * UGSK * p_y(0)}$
2. if $y \in$ non-leaf node

(a) Find the share value of all child nodes (ch) first, then apply Lagrange interpolation formula to compute the share value of node y.

(b) if $\forall ch \in y;\ s_y = \perp$ then $s_y = \perp$

else

    i. Let $j = index(ch)$ and $\delta'_y = index(ch): ch \in \delta_y$.

    ii. $s_y = \prod_{ch \in \delta_y} (sv_{ch})^{\triangle_{j,\delta'_y}(0)} = \prod_{ch \in \delta_y} \left(\varrho(\mathfrak{g},\mathfrak{g})^{\alpha * UGSK * p_{ch}(0)}\right)^{\triangle_{j,\delta'_y}(0)} =$

    $\prod_{ch \in \delta_y} \left(\varrho(\mathfrak{g},\mathfrak{g})^{\alpha * UGSK * p_{parent(ch)}(index(ch))}\right)^{\triangle_{j,\delta'_y}(0)} =$

    $\prod_{ch \in \delta_y} \varrho(\mathfrak{g},\mathfrak{g})^{\alpha * UGSK * p_{ch}(j) \triangle_{j,\delta'_y}(0)} = \varrho(\mathfrak{g},\mathfrak{g})^{\alpha * UGSK * p_y(0)}$

3. if $s_{r_o} = \perp$ then Return $PDCT = \perp$

else

(a) $A = \frac{\varrho(US_1 * US_2, CT_2)}{s_{r_o}} = \frac{\varrho(\mathfrak{g}^{(\kappa * UGSK + \alpha * UGSK - uid)} \cdot \mathfrak{g}^{uid}, \mathfrak{g}^\beta)}{\varrho(\mathfrak{g},\mathfrak{g})^{\alpha * UGSK * \beta}} =$

$\frac{\varrho(\mathfrak{g},\mathfrak{g})^{\kappa * UGSK * \beta} \cdot \varrho(\mathfrak{g},\mathfrak{g})^{\alpha * UGSK * \beta}}{\varrho(\mathfrak{g},\mathfrak{g})^{\alpha * UGSK * \beta}} = \varrho(\mathfrak{g},\mathfrak{g})^{\kappa * UGSK * \beta}$

(b) Return $PDCT = (A, CT_1)$

After partial decryption, the PS sends PDCT to DU.

5.8 Decryption

After receiving the PDCT from PS, the DU decrypts the PDCT with the UGSK using the decryption algorithm. if $PDCT = \perp$ then this algorithm returns $\perp$ otherwise it obtains message as $M = \frac{CT_1}{(A)^{1/UGSK}} = \frac{M \cdot \varrho(\mathfrak{g},\mathfrak{g})^{\kappa\beta}}{(\varrho(\mathfrak{g},\mathfrak{g})^{\kappa * UGSK * \beta})^{1/UGSK}} = \frac{M \cdot \varrho(\mathfrak{g},\mathfrak{g})^{\kappa\beta}}{\varrho(\mathfrak{g},\mathfrak{g})^{\kappa * UGSK * \beta * 1/UGSK}} = \frac{M \cdot \varrho(\mathfrak{g},\mathfrak{g})^{\kappa\beta}}{\varrho(\mathfrak{g},\mathfrak{g})^{\kappa\beta}}$. If only the δ of DU satisfies τ in the ciphertext, then the DU obtains the exact plaintext or message M.

6 Security Analysis

Here, we show the proof for our scheme is secure and it resists the Chosen-plaintext attack (CPA) and secret key forging attack.

Theorem 1. *The PPT adversary ($\mathcal{A}$) has a negligible advantage in selectively breaching our T-CP-ABE scheme if DBDH assumption holds.*

Proof: *To prove our scheme is safe against CPA, we design a security game. The challenger (ς) and ($\mathcal{A}$) are players in this game. The ς chooses random numbers $a, b, c \in Z_p$ and $\mathfrak{R} \in G_t$. The ς tosses the binary coin σ. It outputs either $\sigma = 0$ or $\sigma = 1$. If $\sigma = 0$, then compute $\mathbb{Z} = \varrho(\mathfrak{g},\mathfrak{g})^{abc}$ else assign $\mathfrak{R}$ into $\mathbb{Z}$. Now the DBDH tuple is $(\mathfrak{g}, A, B, C, \mathbb{Z}) = (\mathfrak{g}, \mathfrak{g}^a, \mathfrak{g}^b, \mathfrak{g}^c, \mathbb{Z})$. In this security game, the challenger solves the Decisional Bilinear Diffie-Hellman assumption with advantages of $\frac{\epsilon}{2}$ if $\mathcal{A}$ has an negligible advantage ϵ to selectively breach T-CP-ABE scheme.*

First, the ς sends the PK to $\mathcal{A}$ and also sends US^* (secret key). Then, $\mathcal{A}$ chooses two identical size messages M_0, M_1 and are gives it to the challenger. In turn, challenger sends the ciphertext CT^* for a plaintext M_σ with $\sigma \in \{0,1\}$

to the adversary. Now adversary must guess the exact σ value. If the guess is correct, then $\mathcal{A}$ advantage is ϵ. Thus $Pr[\sigma = \sigma' | \mathbb{Z} = \varrho(\mathfrak{g}, \mathfrak{g})^{abc}] = \frac{1}{2} + \epsilon$. If the guess is not correct, then $\mathcal{A}$ disadvantage is $Pr[\sigma \neq \sigma' | \mathbb{Z} = \mathfrak{R}] = \frac{1}{2}$. Now the advantage of ς to solve the Diffie-Hellman assumption is $\frac{1}{2} Pr[\sigma \neq \sigma' | \sigma = 1] + \frac{1}{2} Pr[\sigma = \sigma' | \sigma = 0] - \frac{1}{2} = \frac{\epsilon}{2}$. Therefore adversary $(\mathcal{A})$ has a negligible advantage in selectively breaching our T-CP-ABE scheme if DBDH assumption holds.

Theorem 2. *Our scheme is secure against secret key forging attack.*

Proof: To breach the tracing, the malicious DU may try to re-randomize the entire secret key or modify any part of the secret key and sell it to the others. To avoid above mentioned breach, our scheme has the key sanity check which ensures the secret key is well-formed. Suppose if the entire secret key is re-randomized or the US_1 parameter is modified then we will not get the exact hash value $\mathfrak{g}^{-H(US_1)}$ during keysanitycheck algorithm step (1). If we do not get the exact hash value then $(US_2 * \mathfrak{g}^{-H(US_1)})$ will not give $\mathfrak{g}^{uid}$ for us to get the correct A value. Similarly if other parameters US_2, US_3 are randomized or modified, then we will not get exact A value as $\varrho(\mathfrak{g}, \mathfrak{g})^{\alpha * UGSK}$ because the uid, $\kappa * UGSK$ values in the randomized or modified US_2, US_3 will not match with the uid, $\kappa * UGSK$ values in the US_1. In the other hand, US_{att} is modified then step (2) in keysanitycheck algorithm $\varrho(X_{att}, US_{att}) = A$, $\forall att \in \delta$ fails. It is noted from all the above cases, the hash function H, and embedding uid & UGSK into multiple parameter of secret key restricts the forged secret key to pass the keysanitycheck. Thus our scheme is secure against secret key forge, and the trace algorithm provides the valid user identity (uid).

7   Performance Evaluation

Here, we analysis the performance of our scheme and compare it with other existing schemes. Table 2 gives the theoretical computation overhead comparison of various cryptographic operations of our scheme and other traceability schemes. Only pairing and exponentiation operations are considered for comparison because other operations comparatively require negligible computation overhead. Table 2 shows that our scheme achieves better performance in all aspects of computations.

Now, we evaluate our proposed T-CP-ABE scheme through experiments. We find the cost required to perform exponentiation and pairing operations over an elliptic curve with symmetric Type 1 pairing (SS512) with the security level 80-bit using the Charm crypto framework [30]. We used the Intel core i7 processor @ 2.50 GHz, 16 GB RAM machine with windows 10 and python 3.2 software for computing the cost. The computation cost of one exponentiation operation required 0.31 ms and pairing operation required 0.62 ms. Here, we omit the computation cost of multiplication, division and hashing operations because it is negligible [31]. We consider the following assumption for the experiments. Let $\eta_\tau = 20$, $\eta_\delta = 10$, and $\eta_{nl} = 10$. We used a random number from G_t as message for encryption and decryption process. The experimental computation cost of proposed scheme and other traceability schemes are given in Table 3. Table 3 shows that our T-CP-ABE required less computation cost in all the cryptography operations. Specifically, our scheme required only 6.82 ms for encryption and key sanity check which is more efficient than other schemes. Furthermore, our scheme reduces the computation overhead of DU as 0.31 ms by outsourcing all the inflated computations to the PS. Thus, our scheme is more efficient than other traceability schemes.

Table 2. Theoretical computation overhead comparison of traceability schemes

Schemes	Secret key	Encryption	Decryption	Outsourced decryption	Key sanity check
LCW [3]	$(\eta_\delta + 4)t_e$	$(3\eta_\tau + 3)t_e$	$(2\eta_\delta + 1)t_p + \eta_\delta t_e$	-	$(2\eta_\delta + 5)t_p$
NDCWL [4]	$(5\eta_\delta + 4)t_e$	$(5\eta_\tau + 3)t_e$	$(3\eta_\delta + 1)t_p + \eta_\delta t_e$	-	$(4\eta_\delta + 4)t_p$
LZYZ [10]	$(4\eta_\delta + 5)t_e$	$(5\eta_\tau + 3)t_e$	t_e	$(3\eta_\delta + 1)t_p + \eta_\delta t_e$	$(4\eta_\delta + 5)t_p$
NCDW [7]	$(\eta_\delta + 11)t_e$	$(3\eta_\tau + 4)t_e$	$(2\eta_\delta + 2)t_p + \eta_\delta t_e$	-	$(2\eta_\delta + 8)t_p$
T-CP-ABE	$(\eta_\delta + 3)t_e$	$(\eta_\tau + 2)t_e$	t_e	$(\eta_\delta + 2)t_p + \eta_{nl}t_e$	$(\eta_\delta + 1)t_p$

t_p - Computation time of one pairing operation; t_e - Computation time of one exponentiation operation; η_δ - No. of user attributes; η_τ - No. of attributes in access policy; η_{nl} - No. of non-leaf nodes in the access tree

Table 3. Experimental computation cost comparison of traceability schemes

Schemes	Key generation cost (ms)	Encryption cost (ms)	Decryption cost (ms)	Outsourced Decryption cost (ms)	Key sanity check cost (ms)
LCW [3]	4.34	19.53	16.12	-	15.5
NDCWL [4]	16.74	31.0	22.32	-	27.28
LZYZ [10]	13.95	31.0	0.31	22.32	27.9
NCDW [7]	6.51	19.84	16.74	-	17.36
T-CP-ABE	4.03	6.82	0.31	10.54	6.82

8   Conclusion

In this paper, we proposed a traceable CP-ABE for outsourced big data in cloud storage. Our scheme achieved the dynamic traceability in CP-ABE schemes for the first time. Our scheme traced the user identity from the user secret key during the outsourced decryption process and stores it into the log file, which helps to identify the malicious users who abnormally access the data. Our scheme also achieved the efficient key sanity check which helps to ensure the secret key is well-formed. We optimized our scheme to cut down the storage overhead for traceability and also cut down the computation overhead for secret key generation, encryption, and decryption process. Especially, the data user's computation overhead is reduced by outsourcing all inflated pairing computations to the proxy server which helps to access the data using resource constraint devices. Since our T-CP-ABE has advantage of both security and efficiency, it would be appropriate for big data in cloud storage. In the future, we would like to extend this paper by augmenting automatic malicious user revocation. In addition, there is an open security challenge when the authority's trust assumption is withdrawn because authority knows the secret key of all users.

References

1. Takabi, H., Joshi, J.B., Ahn, G.J.: Security and privacy challenges in cloud computing environments. IEEE Secur. Priv. **8**(6), 24–31 (2010)
2. Gupta, B., Agrawal, D.P., Yamaguchi, S.: Handbook of Research on Modern Cryptographic Solutions for Computer and Cyber Security. IGI Global, Hershey (2016)
3. Liu, Z., Cao, Z., Wong, D.S.: White-box traceable ciphertext-policy attribute-based encryption supporting any monotone access structures. IEEE Trans. Inf. Forensics Secur. **8**(1), 76–88 (2013)
4. Ning, J., Dong, X., Cao, Z., Wei, L., Lin, X.: White-box traceable ciphertext-policy attribute-based encryption supporting flexible attributes. IEEE Trans. Inf. Forensics Secur. **10**(6), 1274–1288 (2015)
5. Ning, J., Cao, Z., Dong, X., Wei, L., Lin, X.: Large universe ciphertext-policy attribute-based encryption with white-box traceability. In: European Symposium on Research in Computer Security, pp. 55–72. Springer (2014)
6. Zhang, K., Li, H., Ma, J., Liu, X.: Efficient large-universe multi-authority ciphertext-policy attribute-based encryption with white-box traceability. Sci. China Inf. Sci. **61**(3), 032102 (2018)
7. Ning, J., Cao, Z., Dong, X., Wei, L.: White-box traceable CP-ABE for cloud storage service: how to catch people leaking their access credentials effectively. IEEE Trans. Dependable Secur. Comput. **15**(5), 883–897 (2018)
8. Ning, J., Dong, X., Cao, Z., Wei, L.: Accountable authority ciphertext-policy attribute-based encryption with white-box traceability and public auditing in the cloud. In: European Symposium on Research in Computer Security, pp. 270–289. Springer (2015)
9. Ning, J., Cao, Z., Dong, X., Liang, K., Wei, L., Choo, K.K.R.: Cryptcloud+: secure and expressive data access control for cloud storage. IEEE Trans. Serv. Comput. (2018). https://doi.org/10.1109/TSC.2018.2791538
10. Li, Q., Zhu, H., Ying, Z., Zhang, T.: Traceable ciphertext-policy attribute-based encryption with verifiable outsourced decryption in ehealth cloud. Wirel. Commun. Mob. Comput. **2018**, Article ID 1701675 (2018). https://doi.org/10.1155/2018/1701675
11. Bethencourt, J., Sahai, A., Waters, B.: Ciphertext-policy attribute-based encryption. In: IEEE Symposium on Security and Privacy, pp. 321–334. IEEE (2007)
12. Cheung, L., Newport, C.: Provably secure ciphertext policy ABE. In: Proceedings of the 14th ACM Conference on Computer and Communications Security, pp. 456–465. ACM (2007)

13. Waters, B.: Ciphertext-policy attribute-based encryption: an expressive, efficient, and provably secure realization. In: International Workshop on Public Key Cryptography, pp. 53–70. Springer (2011)

14. Li, L., Gu, T., Chang, L., Xu, Z., Liu, Y., Qian, J.: A ciphertext-policy attribute-based encryption based on an ordered binary decision diagram. IEEE Access **5**, 1137–1145 (2017)

15. Jiang, Y., Susilo, W., Mu, Y., Guo, F.: Flexible ciphertext-policy attribute-based encryption supporting AND-gate and threshold with short ciphertexts. Int. J. Inf. Secur. **17**(4), 463–475 (2018)

16. Zhang, Y., Chen, X., Li, J., Wong, D.S., Li, H., You, I.: Ensuring attribute privacy protection and fast decryption for outsourced data security in mobile cloud computing. Inf. Sci. **379**, 42–61 (2017)

17. Kumar, P.P., Kumar, P.S., Alphonse, P.: An efficient ciphertext policy-attribute based encryption for big data access control in cloud computing. In: Ninth International Conference on Advanced Computing (ICoAC), pp. 114–120. IEEE (2017)

18. Li, J., Li, X., Wang, L., He, D., Ahmad, H., Niu, X.: Fuzzy encryption in cloud computation: efficient verifiable outsourced attribute-based encryption. Soft Comput. **22**(3), 707–714 (2018)

19. Premkamal, P.K., Pasupuleti, S.K., Alphonse, P.: A new verifiable outsourced ciphertext-policy attribute based encryption for big data privacy and access control in cloud. J. Ambient Intell. Human Comput. 1–15 (2018). https://doi.org/10.1007/s12652-018-0967-0

20. Yang, K., Han, Q., Li, H., Zheng, K., Su, Z., Shen, X.: An efficient and fine-grained big data access control scheme with privacy-preserving policy. IEEE Internet Things J. **4**(2), 563–571 (2017)

21. Liu, Z., Jiang, Z.L., Wang, X., Yiu, S.: Practical attribute-based encryption. J. Netw. Comput. Appl. **108**(C), 112–123 (2018)

22. Teng, W., Yang, G., Xiang, Y., Zhang, T., Wang, D.: Attribute-based access control with constant-size ciphertext in cloud computing. IEEE Trans. Cloud Comput. **5**(4), 617–627 (2017)

23. Susilo, W., Yang, G., Guo, F., Huang, Q.: Constant-size ciphertexts in threshold attribute-based encryption without dummy attributes. Inf. Sci. **429**, 349–360 (2018)

24. Odelu, V., Das, A.K., Rao, Y.S., Kumari, S., Khan, M.K., Choo, K.K.R.: Pairing-based CP-ABE with constant-size ciphertexts and secret keys for cloud environment. Comput. Stand. Interfaces **54**, 3–9 (2017)

25. Zhang, Y., Zheng, D., Chen, X., Li, J., Li, H.: Efficient attribute-based data sharing in mobile clouds. Pervasive Mob. Comput. **28**, 135–149 (2016)

26. Li, J., Yao, W., Han, J., Zhang, Y., Shen, J.: User collusion avoidance CP-ABE with efficient attribute revocation for cloud storage. IEEE Syst. J. **12**(2), 1767–1777 (2018)

27. Li, J., Ren, K., Kim, K.: A2BE: accountable attribute-based encryption for abuse free access control. IACR Cryptology ePrint Archive 2009, 118 (2009)

28. Li, J., Huang, Q., Chen, X., Chow, S.S., Wong, D.S., Xie, D.: Multi-authority ciphertext-policy attribute-based encryption with accountability. In: Proceedings of the 6th ACM Symposium on Information, Computer and Communications Security, pp. 386–390. ACM (2011)

29. Katz, J., Schröder, D.: Tracing insider attacks in the context of predicate encryption schemes. In: ACITA 2011 (2011)

30. Akinyele, J.A., Garman, C., Miers, I., Pagano, M.W., Rushanan, M., Green, M., Rubin, A.D.: Charm: a framework for rapidly prototyping cryptosystems. J. Cryptogr. Eng. **3**(2), 111–128 (2013)
31. Li, H., Lin, X., Yang, H., Liang, X., Lu, R., Shen, X.: EPPDR: an efficient privacy-preserving demand response scheme with adaptive key evolution in smart grid. IEEE Trans. Parallel Distrib. Syst. **25**(8), 2053–2064 (2014)

A Novel Solution for Virtual Server on the Data Consistency Maintenance in Cloud Storage Systems

Van Thang Doan[1], Vo Quang Hoang Khang[1], Ha Huy Cuong Nguyen[2(✉)], Cong Phap Huynh[2], and Phayung Meesad[3]

[1] Industrial University of Ho Chi Minh, Ho Chi Minh City, Vietnam
vanthangdn@gmail.com, vqhkhang@gmail.com
[2] The University of Danang - College of Information Technology, Da Nang, Vietnam
nguyenhahuycuong@gmail.com, hcphap@gmail.com
[3] Faculty of Information Technology,
King Mongkut's University of Technology North Bangkok, Bangkok 10800, Thailand
pym@kmutnb.ac.th

Abstract. Currently, systems P2P cloud is built on the resources of computing, to implement features such as storage and communication, it is possible to find these cloud systems in smart homes. The systems P2P cloud must also ensure that the salient features of clouds—on-demand resource provisioning, elasticity and measured service—are maintained. In this era, several organizations are storing their data on P2P cloud storage in order to meet the requirements of efficient operation like stability, scalability, and availability of services. Data replication services in cloud storage systems are there to improve performance. In this context, the requirements for ensuring data consistency became increasingly important. In this paper, we propose a new technical solution the data consistency maintenance in Cloud Storage Systems. In this manuscript, we have shown that the proposed solution yields the results for the schema which ensures consistent data on costs and latency. In this paper, we use the Open Stack tool, which incorporates a proposed algorithm Balancing Consistency Availability On System Physical Machine for maintaining data consistency in Cloud Storage Systems.

Keywords: Virtual machine · Besteffort · Distributed environments ·
Cloud Storage Systems · Maintaining data consistency ·
Cloud computing

1 Introduction

Cloud Storage Systems are widely used because of their highly efficient response to distributed systems. The increasing size of the data is the main concern of our society. To maintain the storage file efficiently, we have focused in this manuscript. To minimize the time dispersed by the user, cloud allows easy

© Springer Nature Switzerland AG 2020
P. Boonyopakorn et al. (Eds.): IC2IT 2019, AISC 936, pp. 227–234, 2020.
https://doi.org/10.1007/978-3-030-19861-9_22

deployment of services to develop applications as required by users on the basis of separate physical data centers. Data replication is an effective service (response time data, high availability of data, system performance) of the system storage applications in Cloud environments. The reason is that users can go directly to the copy in data centers (Data Center - DC) nearest copy center at the other ... (*an example is shown below*). However, data centers have problems such as load balancing, fault tolerance, redundancy, or lack of system resources (memory, processing). Moreover, there are many replicas $(R_1, R_2, ...)$ at the physical data centers (Fig. 1) will make more difficult to ensure data consistency.

Therefore, in recent times, solutions using virtual server are interested in research, application. That built on the physical server platform can overcome these problems thus improving the efficiency of the scheme to ensure data consistency. Each solution to these problems by virtual servers can be considered as an approach to improve an efficiency consistency maintenance scheme for cloud storage systems. We use the virtual server solutions to ensure data consistency by Amazon Dynamo.

Dynamo is an eventually-consistent system. Updates are asynchronously propagated to replicas. However, in this paper, we propose solutions for virtual servers to work more efficiently, thereby improving the efficiency of the schema to ensure consistent data. Specifically, with the method for predicting the distributed systems with least completion time which is used to make a complex commercial decision in resource allocation and scheduling. The method of data consistency of maintenance for two-way search algorithms cloud storage. It has brought reliability and efficiency in deploying applications in the cloud.

In this manuscript, the authors propose a novel solution that uses virtual servers to update the replicas. Specific contributions are: (1) It is related resource allocation solution which enables detection and handling deadlocks for virtual servers; (2) Conduct experiments and compares to prove the effectiveness of the new proposal. The work is organized in the following way: in part 2, we have introduced the related works and background study. In Sect. 3, we initiated model virtual server on the Data Consistency Maintenance in Cloud Storage Systems; Sect. 4 deals with the result analysis. This section concludes with the comments and suggestions for future work.

2   Related Works

This Section deals with the related study and background work. Content distribution as a collective service is available with Open Stack. Environmental templates, Open Stack packages or any type of file, for general infrastructure or for your applications can be distributed through P2P file sharing between peers. The CDT of Subutai (content distribution network) allows you to publish publicly or privately share files using global or proprietary. P2P integrated website and verified content storage services allow safe content distribution on Internet-based peer machines. In this paper, currently studying ways to take a step further, considered providing general content distribution as a distributed file system. In [1], the integration of different file systems that can adapter well to the P2P model like Interplanetary File System (IPFS). In systems Peer-to-Peer based Page Rank [2], for distributed startup teams and SME's with multiple branch offices. Enable them with common back-office and productivity applications they need. Native built-in social cloud services and the AppScale based PaaS layer can be utilized to quickly build your ultimate application on Open Stack. In [3], Nakashima proposed an approach to build a tree by recursively selecting a representative and partitioning method. With this method, it supports only to store information of the tree structure in their sub- space for all the intermediate nodes. Therefore, we can see that most of the papers have not focused on the case study of strong consistency. In cloud storage services, maintenance consistency plays an important role in the context of resource allocation for virtual server. In [4,5] resource allocation for Infrastructure as a Service (IaaS), especially the problem of providing heterogeneous resources. The trend of providing resources to meet and maintain storage time in the cloud, when there is a need to increase and replace new resources while running applications is a complex and urgent issue in scheduling techniques principles in IaaS layer. In this paper, we have shown the solution methodology to update consistency in this above context.

3 Virtual Server on the Data Consistency Maintenance in Cloud Storage Systems

In computing server, several hardware and software resources are available for being configured. Hence allocation of resources should be in an efficient manner so that we may utilize the resources in a optimal way.

Resource Allocation: Over the years, the model cloud computing has drawn the attention of researchers Cloud users around the world. Cloud computing presents a model of resource allocation other than grid or scheduling. Especially, Amazon C2 is capable of allocating smaller computing resources, rather than a few, large requirements. In distribution system traditional, we can see a system does not even include a files that is too connected by a network transport. Slow delay in transactions are an limit but not guess before [3,6,7].

Example 1. A requirement of data consistency.

We have used the platform graph, for the grid platform. We model a collection of heterogeneous resources and the communication links between them as the nodes and edges of an undirected graph. See a example in Fig. 2 with simple platform on Industry 4.0. In Industry 4.0 is indispensable are the following virtual machines provide services. Customers here are farmers who have the farms they may need to store data, retrieve data.

In this paper, the practical implications dictate that designers opt for best-effort availability, thus guaranteeing consistency, and greedy consistency for systems that must guarantee availability. With a pragmatic way to handle the tradeoff is by balancing consistency availability tradeoff in systems.

4   Result Analysis

This section deals with the result analysis and discussion. To do so, the solution includes two steps: Firstly, Creation of structured tree; Secondly, checking if cycle exists in that tree. In the first step multicast to be used in dDT algorithm construction in order to send request to on the nodes in the systems and waiting for replies from them. After this process, we obtain a tree.

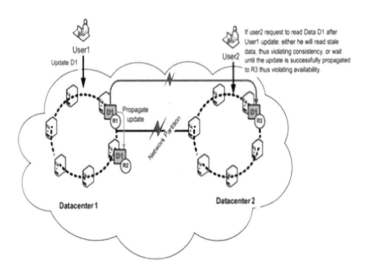

Fig. 1. A requirement of data consistency

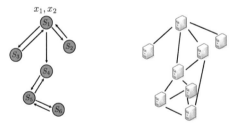

Fig. 2. Example the problem with dissemination tree illustration 2-Ary.

Through Table 1, they apply algorithm request resource with the ability to provide resources. We show the ratio of the number of successful update to the total number of generated replicas.

In this scenario, virtual node replaces fixed physical nodes and this approach incorporates better feasibility, productivity, and load balancing procedures. In a

---

Algorithm 1. Balancing Consistency Availability On System Physical Machine

---

Input: F_i, PM_i, BCA_i, FA_i from IaaS provider i;

Output: Balancing Consistency Availability;

BEGIN

//at the root check $ID_j \in ws_R^n$ // $n = \overline{1,d}$

If has not yet its children N_R^n **Then**

Return R

$ws_j := ws_R^n$

Else if

//q has not yet its children N_R^n

// $ID_j \in ws_R^n$ // $n = \overline{1,d}$

$ws_j := ws_R^n$

Return N_R^n

END

Table 1. Compare the average time between capabilities in resource storage requirements

	VM NOSR	VM SR	VM ID	Start time	Start time	End time
10%	30,56	40,72	20,5	0.1	10	22.22%
20%	40,09	45,21	22,3	0.1	10	20.00%
30%	45,30	45,48	23,18	0.1	32	27.27%
40%	47,54	45,45	24,34	0.1	45	43.75%
50%	45,56	46,01	30,14	0.1	40	39.39%
60%	46,18	47,01	40,01	0.1	20	36.05%
70%	47,02	47,09	43,09	0.1	35	42.86%
80%	48,08	47,56	45,23	0.1	48	44.44%
90%	49,02	48,56	45,67	0.1	60	50.55%
100%	56,78	50,06	60,78	0.1	90	64.86%

ring, a node is assigned with a data item. On the ring, the position is determined by the value. Now the closest node is assigned with the data clockwise. The data can be replicated by the scheduled protocol. In this case, all the nodes stand for equal responsibility. They can easily compute the reference list (Fig. 3).

In traditional distributed storage and cloud computing systems, the insinctive and correct way to handle replicas consistency was to insure a strong consistency state of all replicas in the systems all the systems all the time. Through chart 3, the method reducing the delay of updates propagation as 20% unit time (NR = 1000), respectively (Fig. 4).

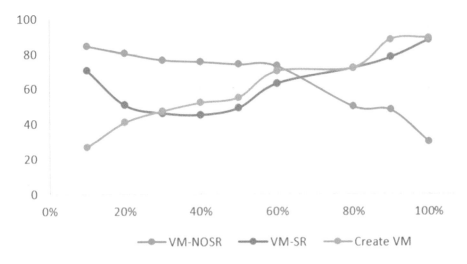

Fig. 3. Impact of the node's join/leave rate on latency

We compare our algorithms with algorithms greedy with detecting and eventual balancing consistency. The first tolerable status success create virtual machine rate 30% (Table 2).

Table 2. Mean attenuation limit with more experiments

	VM NOSR	VM SR	VM ID	Start time	Start time	End time
10%	50,56	60,72	20,5	0.1	5	22.22%
20%	60,09	75,21	22,3	0.1	10	20.00%
30%	65,30	75,48	23,18	0.1	12	27.27%
40%	67,54	75,45	24,34	0.1	15	43.75%
50%	75,56	76,01	30,14	0.1	20	39.39%
60%	76,18	77,01	40,01	0.1	23	36.05%
70%	77,02	77,09	43,09	0.1	25	42.86%
80%	88,08	78,56	45,23	0.1	28	44.44%
90%	89,02	78,56	45,67	0.1	30	50.55%
100%	96,78	80,06	60,78	0.1	35	64.86%

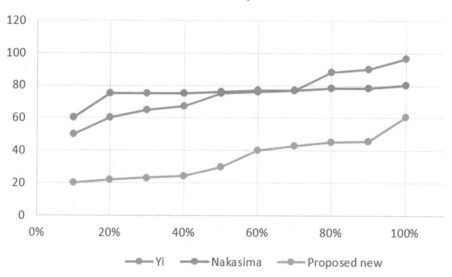

Fig. 4. Impact of the number of tree nodes on latency

5   Conclusions

In the content of the article, we provide Open Stack virtual server solution. The solution concerns the readiness criteria, because it affects the cost of preparing the infrastructure. Virtualization solutions based on Open Stack have great potential to meet the needs of the users in the context of complex and sophisticated intelligent computing systems.

Using a virtual machine is an effective solution that ensures consistent data replication of Cloud Storage Systems. In this paper, the team proposed solutions to improve the performance of virtual servers in allocating system resources; algorithms to prevent, handle deadlock. Experimental results indicate that new proposals are effective. Main issues, the ability to adjust resources can positively impact each user's needs, the solution to ensure quality standards, ensure sufficient resources, storage space, support policies Use when demand increases or decreases. Standalone independence also improves on open resource standards, based on physical server resources.

We have also conducted experiments in distributed environments, some peer-to-peer distributed applications with the ability of data centers to change. Based on empirical evaluation criteria, we propose to bring some positive results.

Through this research, we found that adopting Open Stack based virtual server solutions delivers optimal performance for the distributed resources of specially adapted virtual machine systems. Provide virtual servers for high tech applications in intelligent farming.

References

1. Yazir, Y.O., Matthews, C., Farahbod, R.: Dynamic resource allocation in computing clouds using distributed multiple criteria decision analysis. In: IEEE 3rd International Conference on Cloud Computing, pp. 91–98 (2010)
2. Adami, D., Gabbrielli, A., Giordano, S., Pagano, M., Portaluri, G.: A fuzzy logic approach for resources allocation in cloud data center. In: Proceedings 2015 IEEE Globecom Workshops (GC Wkshps), pp. 1–6 (2015)
3. Nakashima, T., Fujita, S.: Tree-based consistency maintenance scheme for peer-to-peer file sharing systems. In: 2013 First International Symposium on Computing and Networking (CANDAR), pp. 187-193. IEEE (2013)
4. Nguyen, H.H.C., Dang, V.T., Le, V.S.: Technical solutions to resources allocation for distributed virtual machine systems. Int. J. Comput. Sci. Inf. Secur. **13**(2) (2015)
5. Nguyen, H.H.C., Solanki, V.K., Thang, D.V., Thuy, N.T.: Resource allocation for heterogeneous cloud computing. J. Netw. Protoc. Algorithms **09**(2), 71–84 (2017)
6. Rowstron, A., Druschel, P.: Pastry: scalable, distributed object location and routing for large-scale peer-to-peer systems. In: IFIP/ACM International Conference on Distributed Systems Platforms and Open Distributed Processing, pp. 329–350. Springer, Heidelberg (2001)
7. Bermbach, D., Kuhlenkamp, J.: Consistency in distributed storage systems. In: Networked Systems, pp. 175–189. Springer, Heidelberg (2013)
8. Jin, H., Ibrahim, S., Bell, T., Qi, L., Cao, H., Wu, S., Shi, X.: Tools and technologies for building the clouds. In: Cloud computing: Principles Systems and Applications, pp. 3–20, August 2010
9. Kshemkalyani, A.D., Singhal, M.: Distributed Computing Principles, Algorithms, and Systems. Cambridge University Press, Cambridge (2008)
10. Shen, H., Liu, G., Chandler, H.: Swarm intelligence based file replication and consistency maintenance in structured P2P file sharing systems. IEEE Trans. Comput. **64**, 2953–2967 (2015)
11. Li, Z., Xie, G., Li, Z.: Efficient and scalable consistency maintenance for heterogeneous peer-to-peer systems. IEEE Trans. Parallel Distrib. Syst. **19**, 1695–1708 (2008)
12. Nassermostofi, F.: Toward authentication between familiar Peers in P2P networking systems. In: Proceeding of the 9th GI Conference on Autonomous Systems 2016, pp. 88–103 (2016)
13. Pang, X., Wang, C., Zhang, Y.: A new P2P identity authentication method based on zero-knowledge under hybrid P2P network. TELKOMNIKA Indones. J. Electr. Eng. **11**(10), 6187–6192 (2013)
14. Chen, X., Ren, S., Wang, H.: SCOPE: scalable consistency maintenance in structured P2P systems. In: 24th Annual Joint Conference of the IEEE Computer and Communications Societies, INFOCOM, pp. 1502–1513 (2005)
15. Vishnumurthy, V., Francis, P.: On heterogeneous overlay construction and random node selection in unstructured P2P networks. In: Proceedings of the IEEE INFOCOM (2006)
16. Zhao, B.Y., et al.: Tapestry: a resilient global-scale overlay for service deployment. IEEE J. Sel. Areas Commun. **22**, 41–53 (2004)

Data Integration Patterns in the Context of Enterprise Data Management

Roland Petrasch[(⊠)]

Faculty of Science and Technology, Thammasat University,
Rangsit Campus, Khlong Luang 12120, Pathum Thani, Thailand
roland.petrasch@gmail.com

Abstract. Enterprise Data Management comprises various tasks such as providing the strategies, concepts, infrastructure, and tools for OLTP (Online Transaction Processing) and OLAP (On-line Analytical Processing). One important factor is the data integration so that different IT systems are able to exchange transactional and analytical data in a controlled and consistent manner. In the case of OLTP, different options are possible, e. g. direct coupling of systems that lead to uni- or bi-directional integration. However, when multiple systems have to collaborate, a dedicated enterprise data management system might be considered. A premise for this solution is a normalized, consolidated enterprise data model that covers all data or entity models of the systems that are to be integrated. Data Integration Patterns help the business analyst, data scientist, and other IT experts to discuss, document, and conceptualize such an enterprise data model and the data integration tasks. This paper presents Data Integration Patterns for OLTP in the context of the integration of data-intensive IT systems with the goal to use an enterprise data management system. Because Enterprise Data Management is tightly coupled with Enterprise Integration Management, the Data Integration Patterns presented in this paper are also discussed in conjunction with Enterprise Integration Patterns.

Keywords: EDM · Enterprise data management · OLTP ·
Online Transaction Processing · Data integration · System integration ·
Enterprise data model · Entity model · Data Integration Patterns ·
Enterprise Integration · Enterprise Integration Patterns

1 Introduction

In the era of Digitalization and Industry 4.0 in general and Big Data, IoT, Data Science, and Enterprise Integration in particular, Enterprise Data Management becomes an increasingly relevant topic. Instead of connecting each IT system of an enterprise in a uni- or bilateral manner that leads to a high number of specialized data- and control-flow-related dependencies and a high degree of coupling, a multi-lateral, decoupled, and enterprise-model-based approach can be beneficial in terms of maintainability, vendor-independence, extensibility, and data efficiency. Master Data Management (MDM, [1, 2]) uses a unique (or enterprise) data model in order to provide an operational database and MDM application software that manages all master data across the

© Springer Nature Switzerland AG 2020
P. Boonyopakorn et al. (Eds.): IC2IT 2019, AISC 936, pp. 235–244, 2020.
https://doi.org/10.1007/978-3-030-19861-9_23

different IT systems [3]. In this context, the term operational database refers to OLTP (Online Transaction Processing) and not to OLAP (On-line Analytical Processing). MDM can be considered as being a part or aspect of Enterprise Data Management (EDM, [4]) with a focus on data-oriented topics. However, EDM also needs to take business-processes into account, because process-oriented aspects like the context (transactional, non-transactional), access rights, response time, e. g. real-time data availability, are relevant for data management activities. Therefore, in the context of this paper, the data integration patterns are to be discussed in the context of system integration (or enterprise integration patterns to be more specific).

Data integration (DI) can (but does not have) to use a central and dedicated enterprise data management system (EDMS): While there are advantages like a unique and semantically rich data model, disadvantages exist, e. g. additional redundancy of data elements (in operational systems and in the EDMS). It is not possible to postulate that an EDMS is needed independently of the specific and individual requirements of an enterprise, because it depends for instance on the type and amount of data stored in the operational systems and the need for a consolidated and vendor-neutral unique data model.

This paper focuses on data integration, especially DI patterns for data mapping [5] of data model elements in the context of OLTP where a central operational enterprise data management system is to be used. The patterns are on an operational, practical, and concrete level – in contrast to high-level data integration patterns like the Bi-directional Sync. pattern (s. Sect. 3 "Related Work").

2   Operational Data Model and Enterprise Data Model Mapping

Data models of operational systems, e. g. ERP (Enterprise Resource Planning), CRM (Customer Relationship Management), or PLM (Product Life-cycle Management), are often implemented in the form of a relational data model stored in a relational database management system [7]. Relational data models (described in 1969 by Edgar F. Codd) have drawbacks [6], e. g. poor semantics. Entity-relationship models (presented by Peter Chen in 1976) lead to improvements on the semantic level.

Object-oriented data models [8] offer concepts like inheritance, encapsulation (fields and methods), and with standards like OMG UML [9] and OCL [10] a formal and common language is available that describes not only the data elements, but also the constraints (business rules) and the behavior of entities (classes).

Data mapping has various aspects like the mapping type (field, entity etc.), the multiplicity (1:n, n:1 etc.), direction (uni- or bi-directional), transactional or non-transactional, and data conversion (aggregate, format, encode etc.).

Mapping of elements of a conceptual (semantic) data model can be of different types:

- Field Mapping: This is the most basic type of mapping where a single or multiple fields are mapped directly to another field or another group of fields.
- Entity Mapping: A complete entity is mapped to another entity which implies that either the field mapping for each field of an entity is explicitly specified or can be determined automatically.
- Inheritance Mapping: Here, two cases have to be distinguished: a) Multiple relations (of a relational data model) are to be mapped to an object-oriented inheritance tree or b) an inheritance trees must be mapped into a different inheritance tree (or a set of entities that have association instead of inheritance relationships).
- Association Mapping: A mapping of an association or relationship may lead to dedicated entities (or associations) in the target model.

Depending on the direction of the mapping, a many-to-one (n:1) mapping on the entity level means that multiple entities are merged into one entity, and a one-to-many (1:n) mapping on the entity level means that one entity is split into multiple entities. The theoretical case of a many-to-many (n:m) mapping is only considered in connection with inheritance (see example in Sect. 4).

3   Related Work

Data mappings that are described in the context of data warehousing, OLAP or ETL (extract, transform, load), e. g. in [11], do not take object-oriented models and operational data models as the target model into account. The mapping direction is strictly uni-directional in the direction of the data lake or data warehouse. Mapping patterns are not formally described. Other publications focus on other aspects like data distribution [12] or use specific data formats like XML [13] or languages like AMN (Abstract Machine Notation) formalism [14].

Publications that focus on data integration patterns describe high-level patterns like Migration, Broadcast, Bi-directional Sync, Correlation, Aggregation [15] or Data Consolidation, Data Federation, Data Propagation, Data virtualization [16], but not on the concrete mapping level as it is described in this paper.

Enterprise Integration in general and Enterprise Integration Patterns (EIP) in particular focus on message-oriented middleware (MoM) components and communication [17], but do not take data model elements and their mapping into account. The same applies to standards like OMG Profile and Interchange Models for Enterprise Application Integration (EAI, [18]).

Tools like "erwin MM" (Mapping Manager) [19] or "Webratio" [20] do not use data integration patterns and focus on specific cases like data model and the web model mapping, respectively.

4 Data Integration Patterns

Analog to the classical software design patterns [21], the patterns presented here should provide solutions for commonly occurring problems within the context of data mappings for data integration (DI patterns).

For the description of patterns a template that contains the attributes like the pattern name can be used [22]. The following example (see Table 1) describes the data mapping pattern "Relations to Inheritance Tree" that occurs when a group of relations (tables) in a relational data model (source model) is to be mapped to an object-oriented structure that represents an inheritance tree consisting of classes (target model). The attribute *type* specifies the aspects mentioned in Sect. 2, e. g. mapping type, multiplicity, direction.

Table 1. Example Data Integration Pattern: Data Mapping for "Relations to Inheritance Tree".

Attribute name	Function
Name	**Relations to Inheritance Tree**
Type	Entity Mapping, 1:n or n:m, uni-/bi-directional, transactional
Objective	A group of relations (tables) in a relational data model (source model) is to be mapped to an object-oriented structure that represents an inheritance tree (target)
Problem	A group of relations represent semantically an inheritance tree and the target model must describe the inheritance structure
Forces	A thorough analysis of the kind of inheritance is necessary. Sub-classes can be overlapping. The Liskov principle [23] is to be taken into account.
Context	Legacy systems often contain one or more relations with redundant attributes that can be mapped into a inheritance tree.
Solution	This pattern can be used when redundant attributes in relations exist and the semantics of the relations allow the mapping to super- and sub-classes.
Results	The advantage of this pattern is that the resulting object-oriented inheritance tree is semantically rich and free of redundant attributes.
Comments	The 1:n case occurs when on relation is to be split into multiple entities that have inheritance relationships

The complete pattern catalog consists of more than 30 patterns; to describe all of them would go beyond the scope of this paper. Therefore, a list of selected DI patterns for data model mappings of relations of a relational data model to elements of an object-oriented class model is presented here. The patterns are quite self-explanatory (see Table 2), but a short description is also provided.

Table 2. Data Integration Patterns: List of selected pattern for Data Mapping

Pattern Name	Icon	Description
Relations-to-Inheritance-Tree	R △	A group of relations is mapped to an object-oriented inheritance tree.
Split Field	F⟨F F	A single field that contains multiple data elements is split into separate fields.
Aggregate Field	F▶F	The content of a field (source model) is to be aggregated. The value is to be used in another entity of the target model.
Relation-to-Class	R↔C	A relation is mapped to a class. This implies that field mappings exists.
Relations-to-Association-Class	R↔C	3 relations with 1:n and n:1 relationships are mapped into 2 classes (with an n:m association) and one association class (object-oriented data model).
Extract-Fields-to-Class	F F C	A group of fields in one or more relations become a separate class in the target model.
Convert-Field-to-Enumeration	F⊢E	The values of a field of a relation are used for the creation of an enumeration.
Merge Fields	F F F	Two or more fields of a relation that are equivalent lead to one single field in a class (with multiple getter methods).
Field-to-Field	F⊢F	A field of a relation of the source model maps to a field of an entity of the target model.
Relationship-to-Associaton	R C	A relationship between two relations leads to an association between classes.

Patterns can be specified for all kinds of source and target models, e. g. the source data model can be an object-oriented class model and the target data model a multi-dimensional data-warehouse data model with data cubes.

A grouping or taxonomy for the patterns should be provided for ease-of-use, e. g. groups *relational-to-object-oriented* and *relational-to-relational* for data model integration where the source model is a relational data model and the target models are object-oriented and relational, respectively. The next group level consists of the mapping element, e. g. field, relation (or class), relationship (or association).

Another aspect for the DI patterns are constraints or rules. This topics will be explained further in connection with an example in the next section.

5 Application of the Date Integration Patterns

An example for the application of Data Integration Patterns is shown in Fig. 1. The following patterns are applied to a relational data model (source model on the left side) and an object-oriented class model (target model on the right side):

- Relations-to-Inheritance-Tree: The relation Account is mapped to the inheritance tree (classes `Customer` and `Person`). The mapping model element has the identifier `AccountToPersonCustomerMapping`.
- Relationship-to-Association: The relationship between `Account` and `Order` (identified via PK-FK) is mapped to an association between the class Customer and `SalesOrder`.
- Relation-to-Class: The relation `Order` has a mapping to the class `SalesOrder` so that the complete entity is mapped. Mapping on the field level must be defined or can be automatically determined if the field identifier are similar or identical.
- Aggregate-Field: A mapping for field values of rows in the relation `OrderLineItem` to a field value in the class `SalesOrder` is defined, i.e. the sum of the prices of the order line items of the source data model is calculated and mapped to the field `total` in the target model class `SalesOrder`.

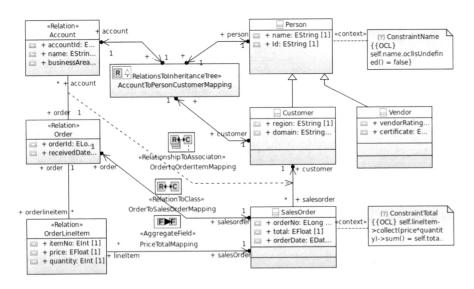

Fig. 1. Example for the application of Data Integration Patterns

Figure 1 also exemplifies the use of constraints used to ensure data integrity and consistency as well as to provide a mechanism for calculations, e. g. rules that describes how a field value is to be determined based on values of another field. The class `SalesOrder` for example has a field `total` that is calculated (sum of all order line items: `price * quantity`).

Another example is the consistency rule that a person's name must not be null. In this case, OCL (Object Constraint Language) is used, but other languages like JPQL (Java Persistence Query Language) or OQL (Object Query Language) could also be applied.

For each DI pattern a sample implementation on the code level should be provided, so that a mapping tool developer has a better picture about the details of the pattern and how to use it for his or her own project. Figure 2 shows the Java code for the Data Integration Pattern "AggregateField" that was used in the model example (s. Figure 1). The `priceTotalMapping` method provides the mapping for a field in the target data model class total in the class `SalesOrder`. A test method is also provided to show how to call the mapping method.

```
/**
 * test priceTotalMapping
 */
public void testpriceTotalMapping() {

    List<OrderLineItem> orderLineItems = new ArrayList<>();
    SalesOrder salesOrder = new SalesOrder();
    BinaryOperator<Float> calcTotal = (price, quantity) -> price * quantity;

    // create some order line otems
    // create a sales order
    priceTotalMapping(orderLineItems, salesOrder, calcTotal); // call mapping function
}

/**
 * priceTotalMapping
 *
 * @param orderLineItems data element from source data model
 * @param salesOrder data element from traget data model
 * @param calculateTotal Operator for calculate line total
 */
public void priceTotalMapping(List<OrderLineItem> orderLineItems, SalesOrder salesOrder,
                              BinaryOperator<Float> calculateTotal) {

    salesOrder.setTotal(orderLineItems.stream()
            .mapToDouble(item -> calculateTotal.apply(item.getPrice(), item.getQuantity()))
            .sum());
}
```

Fig. 2. Sample implementation for the Data Integration Pattern "AggregateField"

6 Data Integration Patterns in the Context of Enterprise Integration Patterns

The complete set of Enterprise Integration patterns (EIP, [17]) cannot be discussed here, but at least two examples for the combination of the patterns is to be provided in order to demonstrate the rationale of the idea to use these patterns together in the context of system integration projects.

Firstly, the EIP "Canonical Data Model" (s. Figure 3) suggests exactly what is called "Enterprise Data Model" in the context of this paper: The usage of a "Data Model that is independent from any specific application" [17]. Translators (see

Message Translator Pattern) are nothing else than data model mappers that "translate" data elements from a source model to a target model (EDM) and the other way around: The EDM can also be used as a source of information.

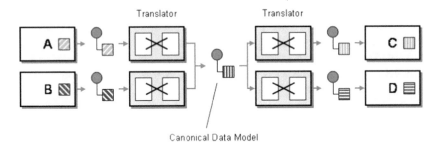

Fig. 3. EIP "Canonical Data Model" and "Message Translator" [17]

Every translator needs a detailed mapping (or translation) specification. At this point, the DI patterns can help in specifying the details of the translations. Figure 4 demonstrates this with an example: 2 messages (with data-sets of relations A, B, C, and D of a relational data model) are to be transferred into an enterprise data model. In term of EIP, they are translated (Message Translator pattern) and in terms of Data Integration patterns, the two patterns "Relations-to-Inheritance-Tree" and "Relations-to-Class" are applied. The pattern icons of the Data Integration patterns appear at the Translator on the upper right corner.

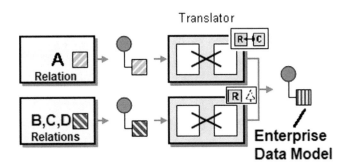

Fig. 4. Combined EIP "Canonical Data Model" and "Message Translator" with Data Integration Patterns "Relations-to-Inheritance-Tree" and "Relations-to-Class"

The second example is the EIP "Aggregator" that presents the idea of a stream of messages that leads to an aggregated message. It remains unclear how exactly this aggregation take place (Fig. 5).

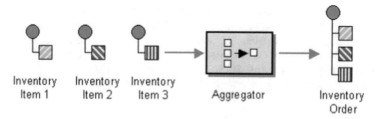

Fig. 5. Enterprise Integration Pattern "Aggregator" [17]

Again, the Data Integration pattern can be helpful to specify the data-oriented details providing answers to questions like "what data elements are used for the aggregation?" and "what data element hold the aggregated value?".

7 Summary

Data integration patterns contribute to a common understanding of how data element mappings can be specified and help experts like data scientists, business analysts, and software developers who need to analyze, document, implement, and test enterprise or data integration applications or components. With formal languages like UML and OCL, it is possible to validate the specification and use them directly for implementation of data integration and data model mapping software components.

Further studies are needed in order to embed the aspect of data integration into enterprise application integration with the help of data integration pattern and enterprise integration patterns, but the results presented here are promising in the sense of reducing the effort for manual coding of data element mappers and converters in the context of system integration projects.

A tool for data integration using the data integration patterns is currently under development so that the approach can be evaluated further on a practical level. A first test was successful. However, the graphical representation has its limits in the case of large data models and a high number of field-mappings. A table view for the specification of mappings is helpful and lets the data modeler specify field-mappings in a more efficient manner.

The model mapping specifications can not only be used for system integration projects in the context of a manual implementation of mapper software components, but also for automatic code generation with transformation and code generation frameworks like OMG QVT and MOFM2T [24, 25].

References

1. Bonnet, P.: Enterprise Data Governance: Reference and Master Data Management Semantic Modeling. Wiley, Hoboken (2013)
2. MDM Alliance Group (MAG). http://www.mdmalliancegroup.com/. Accessed 9 Dec 2019
3. Bonnet, P.: Introduction to MDM Part 1 - A quick IS/IT story Master Data Management. http://docmdm.weebly.com/uploads/1/3/5/4/13543741/introductiontomdmpart1.pdf. Accessed 29 Nov 2019

4. Simon, A.: Modern Enterprise Business Intelligence and Data Management: A Roadmap for IT Directors, Managers, and Architects. Morgan Kaufmann, Burlington (2014)
5. Petrasch, R., Kongkachandra, R.: Towards enterprise data management: data model mapping. In: Conference on Electrical and Computer Engineering 4th IEEE International Women in Engineering (WIE), Pattaya, Thailand, pp. 331–334 (2018)
6. Beynon-Davies, P.: Database systems. Macmillan International Higher Education, p. 115 (2017)
7. Lemahieu, W., vanden Broucke, S., Baesens, B.: Principles of Database Management: The Practical Guide to Storing, Managing and Analyzing Big and Small Data. Cambridge University Press, Cambridge (2018)
8. Papazoglou, M., Spaccapietra, S., Tari, Z.: Advances in Object-oriented Data Modeling. MIT Press, Cambridge (2000)
9. OMG (Object Management Group): OMG Unified Modeling Language (OMG UML), version 2.5, document formal/2015-03-01 (2015)
10. OMG (Object Management Group): Object Constraint Language (OMG OCL) Version 2.4document formal/14-02-03 (2014)
11. Shahbaz, Q.: Data Mapping for Data Warehouse Design. 1st edn. Elsevier (2015)
12. Fresno, J., Torres, Y., Gonzalez-Escribano, A., Llanos, D.R.: An extensible system for multilevel automatic data partition and mapping. In: IEEE Transactions on Parallel and Distributed Systems (2013)
13. Liao, Y., Roman, D., Berre, A.J.: Model-driven rule-based mediation in XML data exchange. In: Proceedings of the First International Workshop on Model-Driven Interoperability, MDI 2010, pp. 89–97. ACM, New York (2010)
14. Manukyan, M.: Canonical Model: Construction Principles. In: iiWAS 2014 Proceedings of the 16th International Conference on Information Integration and Web-based Applications and Services, pp. 320–329. ACM, New York (2014)
15. Mulesoft: Top Five Data Integration Patterns. https://www.mulesoft.com/resources/esb/top-five-data-integration-patterns. Accessed 2 Nov 2018
16. Data Integration and Data Integration Patterns: An Overview. http://www.dbta.com/Editorial/Trends-and-Applications/Data-Integration-and-Data-Integration-Patterns-An-Overview-127077.aspx. Accessed 21 Nov 2018
17. Hohpe, G., Woolf, B.: Enterprise Integration Patterns: Designing, Building, and Deploying Messaging Solutions. Addison-Wesley, Boston (2012)
18. OMG (Object Management Group): UML Profile for Interchange Models for Enterprise Application Integration (EAI). document number: formal/04-03-26 (2004)
19. Cigardi, L.: The Data Model Mapping. https://erwin.com/blog/data-modeling-and-data-mapping-any-data-anywhere/. Accessed 9 Jan 2019
20. McGovern, A.: Data Modeling and Data Mapping: Results from Any Data Anywhere. https://my.webratio.com/learn/learningobject/the-data-model-mapping. Accessed 4 Jan 2019
21. Gamma, E., Helm, R., Johnson, R., Vlissides, J.: Design Patterns: Elements of Reusable Object-Oriented Software. Addison-Wesley, Boston (1995)
22. Dreibelbis, A., Hechler, E., Mathews, B., Oberhofer, M., Sauter, G.: Information service patterns, Part 4: master data management architecture patterns. A pattern taxonomy. https://www.ibm.com/developerworks/data/library/techarticle/dm-0703sauter/index.html. Accessed 30 Sept 2018
23. Bruegge, B., Dutoit, A.H.: Object-oriented Software Engineering: Using UML, Patterns, and Java, p. 317. Prentice Hall (2010)
24. OMG (Object Management Group), "Meta Object Facility 2.0 Query/View/Transformation (QVT)", version 1.3, document number formal/16-06-03 (2016)
25. OMG (Object Management Group): MOF Model to Text Transformation Language (MOFM2T), version 1.0, document number formal/2008-01-16 (2008)

Evolutionary Dynamics of Service Provider Legacy Network Migration to Software Defined IPv6 Network

Babu R. Dawadi[1(✉)], Danda B. Rawat[2], and Shashidhar R. Joshi[1]

[1] Department of Electronics and Computer Engineering, Pulchowk Campus,
Tribhuvan University, Kathmandu, Nepal
{baburd, srjoshi}@ioe.edu.np
[2] Cyber Security and Wireless Networking Innovations Lab, EECS Department,
Howard University, Washington DC, USA
db.rawat@ieee.org

Abstract. Internet Protocol version 6 (IPv6) addressing, Software Defined Network (SDN) and Network Function Virtualization (NFV) are regarded as next generation networking technologies that help to avoid all the existing issues related to address depletion, security, quality of service, scalability and management complexity in the legacy IPv4 networking system. However, the technology migration become one of the central challenges for the stakeholders such as service providers, end users, and regulatory bodies, it becomes more challenging to the internet service providers (ISP) of developing nations due to lack of sufficient cost for migration and limited human resources capable to handle the new networking system. In this paper, we present the economic model of Tier-3 ISPs and present the migration analysis in terms of network utilities to migrate existing network into Software Defined IPv6 (SoDIP6) Network using evolutionary gaming approach. We model the scenario when internet service providers can take a decision for joint migration to SDN enabled IPv6 network and evaluate the migration process using simulation and numerical results. The analysis shows that network migration is an evolutionary process in which taking migration decision is mostly affected by migration status of other interconnected networks, maturity of application and protocols, strength of organization in terms of available budget and technical human resources.

Keywords: IPv6 · SDN · SoDIP6 · Teir-3 ISPs · Network migration · Evolutionary dynamics

1 Introduction

With the advancement on Information and Communication technologies (ICT), we have high volume of networking devices connected to internet that led to the emergence of Internet of Things (IoT). In other words, number of connected devices to the internet is increasing exponentially. Studies have already shown that the legacy IPv4 networks which dominate the current internet, cannot handle the projected number of networked devices and applications. IPv6 has been alternative solution to handle the

© Springer Nature Switzerland AG 2020
P. Boonyopakorn et al. (Eds.): IC2IT 2019, AISC 936, pp. 245–257, 2020.
https://doi.org/10.1007/978-3-030-19861-9_24

issues related to addressing of exponentially increasing networked devices. For the better management and control of the network, SDN is also established in the network world. IPv6 is the network layer protocol invented to avoid the current issues of IPv4 addressing, while SDN has been emerged to avoid the issues of existing network management towards the highly manageable and programmable network by detaching the control plane from each device into the centralized one so called the controller. SDN paradigm is popularly recognized in the data center while its implementation practices in the service provider networks are in progress. IPv6 on the other hand is becoming compulsion for the service providers to migrate their network at the earliest to solve the issues with increasing number of connected devices. Combining the IPv6 and SDN is expected to ease the process of transitioning from legacy to Software Defined IPv6 (SoDIP6) network and improve the overall network performances. The concept of SoDIP6 network is introduced at [1], while it is defined as:

"The future networking infrastructure fully IPv6 operable and controlled/managed by the SDN controller is recognized as a SoDIP6 network".

It is envisaged that the migration to SDN could be prolonged to longer period same as that the target of IPv6 adoption worldwide has been shifting making it undetermined that when the world's network will be IPv6 only. The invention and development of those new concepts and techniques create a bigger challenges [1] in networking for service providers to migrate their existing legacy networks into the SoDIP6 network.

Although IPv6 deployment strategies [2, 3] with the possible approaches of migration have been identified, SDN deployments over the Internet Service Provider's (ISP) and Telecom Service Provider's (Telcos) networks are in the premature stage. In our previous work [4], we verify that the joint migration to Software Defined IPv6 (SoDIP6) network optimizes the migration cost for service providers and hence ISPs/Telcos will be able to properly manage their network migration in a phase with available budget constraints. In this paper, we present the evolutionary process of network migration to newer technologies like SDN and IPv6 jointly from the perspectives of evolutionary gaming approach. Hence, this study is contributory to fairly sustained service providers of developing nations who are in the early stage of their legacy network migration into SDN enabled IPv6 network. So that ISPs can have proper planning and apply strategies while to take the timely decision on network migration based on the available resource limit.

Rest of this paper is organized as follows. Section 2 presents the overview of world-wide ISP network interconnection architecture and economic model of Tier-3 ISPs. The evolutionary dynamics of network migration will be discussed in Sect. 3, while the simulation and analysis of evolutionary process will be presented in Sect. 4. Section 5 presents the related works and the paper will be concluded in Sect. 6.

2   Economic Model of Tier-3 ISPs

In this section, we discuss on the model of ISP network setup, services provided to customers, IP network interconnection and the utilities of ISPs in terms of income and expenditure. We consider a typical Tier-3 ISP for economic model development. Because ISPs of developing nations are basically categorized into Tier-3 ISPs and their network migrations to IPv6 and SDN are in the early stage.

Internet is the world-wide interconnection of network of networks that are highly distributed and heterogeneous. The major sources of revenue generation for an ISP are by providing dedicated network services to the enterprises as well as providing internet connectivity to home users, enterprises and the content providers. ISPs are broadly categorized into Tier-1, Tier-2 and Tier-3 ISPs. Left figure at Fig. 1 presents the world-wide tired ISP network interconnection architecture. Tier-1 ISPs have the major backbone network interconnection and aggregated network infrastructure that are also the root sources of internet contents and services. Tier-2 ISPs are generally regional ISPs that are directly interconnected with Tier-1 ISPs and provide transit services to Tier-3 ISPs and customers. Tier-3 ISPs include local ISPs which directly provide services to end users and the enterprises. ISPs are internetworked in hierarchies in which Tier-2 ISPs are the customer of Tier-1 ISPs while Tier-3 ISPs are the customers of Tier-2 ISPs. The ISPs within the same tier can have private and public interconnection called peering. The interconnected ISPs in the same tier will charge each other according to the amount of traffic exchanged at the point of interconnection on a settlement basis. Figure 1 (right) shows the proposed economic model for Tier-3 ISPs. In our economic model, for the implementation of evolutionary dynamics in migration, we consider the following. (i) Teir-3 ISPs have end customers and enterprises in which they are charged on the basis of bandwidth supplied. They charge other interconnected ISPs (peers) in per port basis only when the traffic volume outgoing exceeds the incoming in the interconnection point. Similarly, some customers are charged based on dedicated services (like MPLS, VPN etc.) provided (ii) SoDIP6 networks are dual stack IPv4 and IPv6 capable under SDN framework where legacy system is available for recovery purpose. Hence, SoDIP6 network is quad stack (IPv4/IPv6 and Legacy/SDN) capable. Tier-1 & Tier-2 ISP networks are SoDIP6 capable, and (iii) Tier-3 ISPs pay to the transit service providers (basically to Tier-2 ISPs) based on per port basis as per bandwidth agreement. We consider extra cost of operation for ISPs that maintain and operate SoDIP6 network, while migration cost is the one time cost to be invested by ISPs during their network migration. In case of legacy IPv4 only ISPs, the extra cost of operation holds cost due to conversion of incoming IPv6 traffic to IPv4 and vice versa.

SDN implementation by the service provider is fairly independent of other interconnected networks and end customers. It deals with the efficient management of networking operations that benefit for efficient services to be provided by the service provider to its customers. Regarding the IPv6 network migration, it is affected by external factors like status of interconnected network migration, demand of end customers and relevant services to them, government regulation as well as registries/registrant management. Migration to SoDIP6 network does not solely depend on the executive head's decision while the application supports, security, quality of service, traffic support by other transit service providers, organizational budget for network migration, availability of technical human resources, expected revenues etc. play vital roles. However, there are several tunneling and translations approaches [3] developed for IPv6, dual stack implementation provides the safe and comfort transition approach for service providers. It helps for smooth transitioning and guarantee uninterruptable services to customers during migration. The coexistence of legacy IPv4 network and IPv6 as well as SDN may span longer period. Hence, we consider that the set of networking devices (switch/routers) are to be migrated into SDN enabled IPv6 network. Dual stack IPv6 network guarantee the backward compatibility so that existing customers running on IPv4 network can be retained and additional new customers can be gained with IPv6 based services. Similarly, for better efficiency, management, control and operation with flexibility & programmability of network can be achieved after migration to SDN together with the features of legacy system helps for better recoverability and fault tolerant in the network. Migrated ISPs can achieve better market growth with gaining extra customers who specially need advance and good quality services. Those services can be provided via SoDIP6 network.

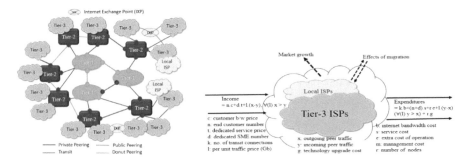

Fig. 1. World-wide ISP network interconnection architecture (left) and Economic model of a Tier-3 ISP (right)

3   Evolutionary Dynamics of Network Migration

The requirements set by Internet Content Providers (ICP), migration status of other transit service providers, requirement of end users and enterprises as well as the government policy and plans affect the decision making for network migration. In this section, we present evolutionary dynamics of network migration so that an ISP in the group of ISPs will decide to move into SoDIP6 network and benefitted with higher utilities after migration. The decision about migrating the IPv4 based network to SDN enabled IPv6 is run by stakeholders such as ISPs, government regulatory bodies and organizations where decisions are not always rational. Thus, we discuss evolutionary dynamics for migration analysis.

We consider the scenario where ISPs that are providing services with legacy IPv4 networking system are categorized into Group 1 and those ISPs offering SoDIP6 network are into Group 2. We assume the scenario of Tier-3 ISPs and follow the assumptions made in Sect. 2 under the proposed economic framework. We assume N number of ISPs are interconnected among with some ISPs have already migrated to SoDIP6 network. The major source of income for an ISP is the internet bandwidth price to be collected from end customers and the revenue collected by providing dedicated services to the enterprises. Similarly, amount of traffic volume exchanged with other interconnected ISPs could contribute to either income in case the outgoing traffic is higher than incoming traffic or will be an expenditure otherwise according to agreement made during the IP interconnection settlement between ISPs. We also incorporate the migration cost in our economic model to measure its importance during migration, but this is one-time cost applicable only after when ISP decides for migration. We expect that ultimately all the ISPs who are in the game play migrate to SoDIP6 network while the probability that one goes for early migration and another decides later will be measured based on the strength of migration determined by the adaptation variable presented in Eq. (3) and the fitness value provided by Eq. (5). In the beginning of the game, we assume Group 2 consists of very few or no ISPs. During the game play, ISPs are migrating to SoDIP6 by setting the strategies based on the advancement on technologies, cost of migration, market growth and future prospects of sustainability. When there are number of ISPs with strategies choosing IPv4 vs the SDN enabled IPv6 network, to simplify the analysis, we consider two ISPs and then we generalize to N ISPs with a division of them into two groups. The list of symbols and their usual meanings are listed in Table 1.

Table 1. List of symbols and their semantics

Symbols	Semantics	Symbols	Semantics
μ_k^4	Utility of kth ISP in legacy IPv4 network	σ_t	Overall adaptation variable value at time step t
p_k^4	Profit of Kth ISP in legacy IPv4 network	σ_c	Adaptation variable based on customer
c_k^4	Cost of Kth ISP operating legacy IPv4 network	σ_p	Adaptation variable based on peer ISPs and traffic
c_k^{s6}	Cost of Kth ISP operating SoDIP6 network	σ_s	Adaptation variable based on human resources, budget and cost
μ_k^{s6}	Utility of kth ISP operating IPv4 with SDN and IPv6	n_t	Total customers of ISP at time step t
p_k^{s6}	Profit of kth ISP operating IPv4 with SDN and IPv6 network	n_{kt}	Number of customers demanding SoDIP6 services at time step t
γ_k	Expected utility of Kth group of ISPs	n_{t-1}	Total customers at previous time step
$X_{4 \to 6}^s$	Number of ISPs having SoDIP6 network	f_{4t}	Fitness function of Legacy IPv4 network at time step t
U_k	Obtained utility of Kth group of ISPs	f_{s6t}	Fitness function of SoDIP6 network at time step t
N_k	Number of ISPs in k^{th} group	n_p	Total number of interconnected peer ISPs
K_k	Number of SoDIP6 ISPs in k^{th} group of ISPs such that $K_k \leq N_k$	n_{p4}	Number of Peer ISPs having Legacy IPv4 only
$ipv6_{in}$	Average volume of IPv6 traffic incoming	$ipv4_{in}$	Average volume of IPv4 traffic incoming
HR_{s6}	Number of technical human resources capable for SoDIP6 network operation	HR_{all}	Number of all technical human resources of an ISP

When both ISPs are running with legacy IPv4 network, then the utility (μ_k^4) for an ISP is:

$$\mu_k^4 = p_k^4 - c_k^4 \tag{1}$$

When one ISP is not migrated and another ISP is migrated to SDN enabled IPv6, utility of non-migrated ISP will be given by Eq. (1) and the utility for SDN enabled IPv6 network will be the income minus the sum of cost associated with IPv6 and SDN operation plus the profit because of the IPv4 to support backward compatibility. This can be expressed as:

$$\mu_k^{s6} = p_k^{s6} - c_k^{s6} + \mu_k^{4s6}(= \mu_k^4) \tag{2}$$

From different studies [4–7], it is realized that the operation cost of SDN and IPv6 network is comparatively less than the existing legacy IPv4 network system i.e. $c_k^{s6} \leq c_k^4$. It means an ISP having SoDIP6 network will have higher payoffs because of its backward compatibility. The possibility of an IPv4 ISP migrates to SoDIP6 network is determined by the readiness status that is measured by the adaptation variable (σ) presented in Eq. (3). Adaption variable for Group 1 ISP at time step t of the game for migration decision is calculated as:

$$
\begin{aligned}
\sigma_{4t} &= \left| \{\sigma_c\} + \{\sigma_p\} + \{\sigma_s\} \right| \\
&= \left| \left\{ \frac{n_{kt}}{n_t} . e^{\frac{(n_{t-1} - n_t)}{n_t}} \right\} + \left\{ \frac{(n_p - n_{p4}).ipv6_{in}}{n_p.(ipv4_{in} + ipv6_{in})} \right\} + \left\{ \frac{HR_{s6t}.B_t}{HR_{all}.C_m} \right\} \right|
\end{aligned} \tag{3}
$$

We consider major three sub-adaptation parameters ($\sigma_c, \sigma_p, and\ \sigma_s$) and their total effects to evaluate $\sigma_{4t}(\sigma)$. For Group 2 ISPs, adaptation variable is not applicable while they are already SoDIP6 capable. Being migrated to advance technology, we don't consider Group 2 ISPs migrate back to legacy IPv4 networking system. σ_c measures the strength towards migration in terms of customer demands for newer technologies with advanced features and change in customer numbers between current and previous timestamp. As σ_p gives the effect on migration in terms of number of directly interconnected ISPs that are SoDIP6 capable and the ratio of IPv6 traffic with the total traffic entering the network. Similarly σ_s provides the effect on migration in terms of migration cost, available budget and the technical human resources. Adaptation variable plays a vital role in the migration decision making. Service providers can evaluate the variable at different time steps like every time step or every round of the game. Based on the past patterns of adaptation values, the migration decision can be taken. Like for example if σ is increasing i.e. $\sigma_{4t-2} < \sigma_{4t-1} < \sigma_{4t}$, the ISP changes the strategy and proceed for migration to SoDIP6 Network. "*How to achieve the increasing σ_{4t}?*" can be evaluated by dissecting the three sub-adaptation parameters σ_c, σ_p and σ_s. With the advancement on new technologies, customer wants faster, efficient and better services and so the customers (n_{kt}) demanding for IPv6 and SDN based services will be increased and if an ISP could not provide the newer services as per the demand, then it may lose the customers. In such case, the customer difference i.e. ($n_{t-1} - n_t$) will be positive resulting to higher σ_c than previous. Similarly, when other ISPs are migrated to SoDIP6, the incoming IPv6 traffic in the Group 1 ISPs would increase and also the number of interconnected SoDIP6 ISPs increases giving to higher σ_p. This in fact measures the external interference in network migration. In the early stage, Group 1 ISPs will have limited human resources capable to operate SoDIP6 network and the limited budget too. ISPs can prepare their human resources according to the increasing strength of σ_p. In evolutionary dynamics, considering the Moran process [9] where the number of populations (i.e. number of ISPs) remains same and relative fitness, each ISP has interactions with its neighboring ISPs. We calculate fitness function with respect to adaptation value and the utilities in which the higher the fitness value is more likely to migrate to SoDIP6. When the game starts and migration populations are randomly chosen at the beginning, for example $X_{4 \to 6}^s (\leq N1)$ members of the Group 1 decided to move to SoDIP6 networks, i.e. $(N1 - X_{4 \to 6}^s)$ members of Group 1 do not have SoDIP6 networks. The expected payoffs of ISP in Group 1 and Group 2 are:

$$\gamma_1 = \frac{\sum_{k=1}^{N1-X_{4 \to 6}^s} \mu_k^4}{N1 - X_{4 \to 6}^s} , \ \gamma_2 = \left\{ \frac{\sum_{k=1}^{N2} \mu_k^{s6}}{N2} + \frac{\sum_{k=1}^{X_{4 \to 6}^s} \mu_k^{4s6}}{X_{4 \to 6}^s} \right\} \tag{4}$$

The fitness of individuals in Group 1 and Group 2 as contributed by payoffs are:

$$f_{4t} = 1 - w + w.\gamma_1 \ and \ f_{s6t} = 1 - w + w.\gamma_2 \tag{5}$$

Where w provides the strength varying from 0 to 1. When $w = 0$, the fitness of both group is 1 meaning that migration does not contribute to higher payoffs and hence migration is not necessary. When $w = 1$ the fitness of both groups depend solely on their payoffs. The ISP once migrated is supposed to be dead from Group 1 and born in Group 2 thus maximizing the profit of group 2. This is simply the birth-death process where same ISP is reproduced after its death. Since, there is no probability that the ISP once moved to Group 2 (i.e. migrated to SoDIP6) will migrate back to Group 1 (legacy IPv4 system). Hence, the probability that the number of SoDIP6 ISPs increases and remains same are calculated using the fitness values [8], if K number of ISPs out of N ISPs have SoDIP6 network at time step t:

$$p_{K,K+1} = \frac{K.f_{s6t}}{K.f_{s6t} + (N-K)f_{4t}} \frac{N-K}{N} \ and \ p_{K,K} = 1 - p_{K,K+1}$$

When $p_{0,0} = p_{N,N} = 1$, this implies that all ISPs apply same strategy in those absorption states. The fixation probability (P_{s6t}) [9] determines scenario that migration to SoDIP6 favors if $P_{s6t} > \frac{1}{N}$, where, $P_{s6t} = \frac{1}{1 + \sum_{k=1}^{N-1} \prod_{j=1}^{k} \frac{f_{4t}(i)}{f_{s6t}(i)}}$

If there is no IPv4 networks, then Eq. (1) gives zero payoff. Thus, the game is formulated for N ISPs with two groups: Group 1 with N_1 legacy IPv4 networks and Group 2 with N_2 SoDIP6 networks with IPv4 compatibility. Note that networks and ISPs are used interchangeably. The utility obtained by the group is shared among the group members based on their proof of stake. Assuming, utility is homogenous for all ISPs, then group utilities are:

$$\gamma_1 = \left(N1 - X_{4\rightarrow6}^s\right).\mu_k^4, \ \gamma_2 = \left(N2 + X_{4\rightarrow6}^s\right)\mu_k^{4s6}$$
$$s.t. \ 0 \leq X_{4\rightarrow6}^s \leq N1 \& N1 + N2 = N \tag{6}$$

Since, when ISPs of Group 1 are migrating and converging towards SoDIP6 network entering into Group 2, then the utility for a given network or ISP k can be expressed as $U_k = \log(\sigma_k + \gamma_k) at \ k = [1, 2]$ where $\log(\sigma_k + \gamma_k)$ is a generic convex function of γ_k for each group k and a typical value of σ_k is 1. Without loss of generality, all members are rewarded equally participating in a group mission and the utility for two groups (IPv4 only i.e., Group 1 and SoDIP6, i.e., Group 2) can be expressed as:

$$U_1 = \log\left(1 + \left(N1 - X_{4\rightarrow6}^s\right).\mu_k^4\right) \quad and \quad U_2 = \log\left(1 + \left(N2 + X_{4\rightarrow6}^s\right).\mu_k^{4s6}\right)$$
$$S.t. \ 0 \leq X_{4\rightarrow6}^s \leq N1 \& N1 + N2 = N \tag{7}$$

From Eq. (7) and its numerical results on Fig. 4, it is seen that the group utility is mostly driven by the number of ISPs migrated to SoDIP6 network in the evolutionary game play. This means, increasing number of ISPs in group 2 increases IPv6 traffic flow within and outside the group leading to increase in incoming IPv6 traffic in group 1 ISPs that adds extra cost for group 1 ISPs to install additional translator device in the border. Similarly, Group 2 utilities increased because of its dual network operating features.

4   Simulations and Analysis

Adaptation to newer technologies are the evolutionary process. Realizing the real scenario, the reproduction and elimination process constitutes the migration of legacy IPv4 network into SoDIP6 network so that number of SoDIP6 capable ISPs increases and Legacy IPv4 capable ISPs decreases. Hence, the normalized utility increases when more ISPs switch from IPv4 to SDN enabled IPv6 network. Table 2 provides assumptions on the initial parameter values and their effects on changing values for this simulation and analysis. We consider a period of 6 years for decision making in the migration process and perform the network migration. The period of 6 years is divided into 24 time steps with each time step consists of 3 months during when ISP calculates the adaptation variable on every subsequent time steps and evaluates the fitness for migration. The initial simulation values of different parameters are set according to the discussion provided in Sect. 3. While, we assume certain fix percentage of increment or decrement on the parameter values provided in Table 2 and perform the analysis based on the numerical results obtained.

At the beginning of the game, ISPs are chosen randomly. However in the random choice, ISPs with higher σ_{4t} can be considered. On the subsequent next steps of game play, ISPs will go for self-decision for migration based on their increasing adaptation value (σ_{4t}). Hence, we consider additional constraints on the evolutionary dynamics to meet the reality that instead of random choice, every time, only Group 1 ISP will be chosen for migration according to the increasing adaptation value for the subsequent next steps of game play. Similarly, Group 2 ISPs can't be chosen for elimination. That

Table 2. Initial assumptions in simulations and analysis

- Total customers = 1200 - Total time period = 6 years - Per time step = 3 months - Legacy IPv4 expected utility = 500 - SoDIP6 expected utility = 700	- IPv4 traffic entered into network: 900 Terabytes - IPv6 traffic entered into network = 300 Terabytes - Total interconnected ISPs = 16 - Peer ISPs migration ratio = 2/3 per time steps.	- Total technical human resources per ISP = 120 - SoDIP6 capable human resources per ISP = 2 - Available budget per ISP = 30,000 USD - Estimated migration cost per ISP: 45,000 USD
σ_c (change per time steps)		σ_p (change per time steps)		σ_s (change per time steps)	
Customer Demand of SoDIP6 Services	Customer Numbers	Incoming IPv6 traffic	Incoming IPv4 traffic	Organizational budget	Migration cost
1%↑	−2%↓	2%↑	5%↑	2%↑	0.2%↑
3%↑	−3%↓	5%↑	−2%↓	3%↑	2%↑
5%↑	−5%↓	6%↑	1.4%↑	5%↑	0.2%↑
0.5%↑	2%↑	10%↑	−2%↓	7%↑	3%↑

means Group 2 ISPs can't migrate back to the category of Group 1 ISPs. Equation (3) provides the overall strength for ISPs to take the migration decision. The sub-adaptation parameters and their effects in migration decision making are separately visualized in Fig. 2 where figure (a) presents the adaptation strengths with respect to customer, figure (b) evaluates the strength from the perspective of traffic engineering and figure (c) provides the strength with respect to available human resources & the migration budget. The different plots show the scenarios of adaptation strengths on rate of change of parameters in the simulations. σ_c at (0.5%↑, 2%↑) indicates the customer demand increased by 0.5% of total customers and 2% increase in total IPv4 customers. It shows the less strength on migration because the ISP is not losing customers however there have been less demands of SoDIP6 network. Losing of customers does not only depend on the demand of new services as there are other causes like reliable services, price of services and meet of quality of service (QoS) requirements as per service the level agreement (SLA). Similarly the plot of σ_c at (5%↑, −5%↓) put higher alarm to ISPs who need to migrate their network into SoDIP6 as early as possible because customer demand of SoDIP6 based services is higher (increased by 5%) and rate of losing customers is also higher (decreased by 5%). In Fig. 2(b), we assume, there is an ISP who has mesh interconnection with other 15 legacy IPv4 ISPs and we simulate the situation of migrating those ISPs into SoDIP6 network and the effects of that migration is plotted over 24 time steps (6 years). 16 ISPs are migrated over 24 time steps, hence, we consider the ISP migration ratio of 16/24 (=2/3). Neighboring ISPs who already migrated to SoDIP6 network may lead to the increase of IPv6 traffic incoming to other interconnected ISPs and hence also leads to decrease in incoming IPv4 traffic. However, higher budget accumulates while delaying the migration, the cost of migration will increase due to the delay because market inflation can't be avoided. Hence, Fig. 2 (c) presents the plots on increasing budget and also increasing migration cost while we consider SoDIP6 network operation capable human resources are increased by 4.16% (1/24) of the existing capable human resources.

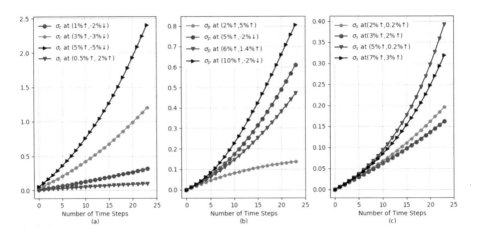

Fig. 2. Simulation results of sub-adaptation parameters for an ISP having legacy IPv4 network at a period of 6 years divided into 24 time steps (3 months per time step)

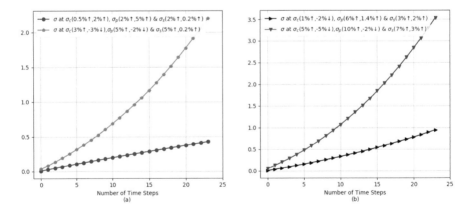

Fig. 3. Plots of $\sigma_{4t}(\sigma)$ at different values of sub-adaptation parameters

Figure 3 shows the cumulative strength of adaptation variable (σ_{4t}) results of sum of all three sub-adaptation variable values at different combinations of their parameters. The pattern of σ_{4t} at Fig. 3(a) indicated by graph label σ_c at $(3\%\uparrow, -3\%\downarrow....)$ and Fig. 3(b) indicated by graph label σ_c at $(5\%\uparrow, -5\%\downarrow....)$ show that it is more likely has to take the migration decision because the customer demand of SoDIP6 is increasing by $(3\%, 5\%)$ with losing of total customers by $(-3\%, -5\%)$, similarly, interconnecting ISPs are migrating to SoDIP6 leading to increase of incoming IPv6

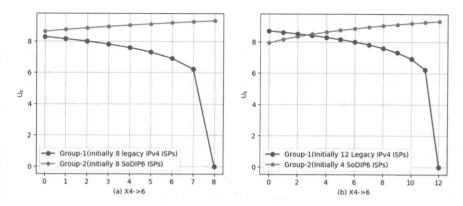

Fig. 4. Plots of utility vs. the number of ISPs transitioning to SoDIP6 network

traffic by (5%, 10%) and decreasing of IPv4 traffic by (−2%, −2%) on every time steps. The yearly budget available is increased by (5%, 7%) shows that ISP having σ_{4t} with graph label σ_c at $(3\%\uparrow, -3\%\downarrow....)$ and σ_c at $(5\%\uparrow, -5\%\downarrow....)$ are stronger than ISP having σ_{4t} with graph label σ_c at $(0.5\%\uparrow, 2\%\uparrow....)$ and σ_c at $(1\%\uparrow, -2\%\downarrow....)$ to take migration decision. In Fig. 4(a) & (b), we present the numerical results of evolutionary dynamics with 8 ISPs in Group 1 and 8 ISPs in Group 2 at (a) and 12 ISPs in Group 1 and 4 ISPs in Group2 at (b) in the beginning. When all IPv4 networks are switched to SDN enabled IPv6 networks, utility for Group 1 is zero during the situation when all ISPs are SoDIP6 capable with higher utilities.

5 Related Works

Authors at [10] presented the game theoretic approach on IPv6 network migration. They focus on the migration possibilities and possible delay on migration in terms of Autonomous Systems (ASes). The concept is more generic and applies to all types of tiered ISPs. The Migration cost is not considered in their migration analysis. Authors in [11] presents game theoretic approach with different mathematical models and scenarios for incremental adoption of IPv6 considering the effects on migration from different stakeholders like service providers, content providers and content users. The issues in joint network migration and the cost benefit analysis were discussed in [1]. Similarly, cost estimation for joint network migration for service providers is presented at [4]. Our focus on this paper is the evolutionary approach of migration for the ease of service providers to take timely decision on network migration. We analyzed the migration process by evaluating the adaptation factor by service providers themselves. Our proposed economic model is also applicable to Tier-2 ISPs having interconnections with other Tier-2 ISPs and have their own customers to provide internet connection and dedicated services.

6   Conclusion

Traditional networks cannot meet the exponential growth of the users because of its static parameters that cannot be changed on the fly and limited space for growing number of subscriptions. To maximize efficiency, scalability, flexibility and manageability of the expanded network and growth of internet users, service providers need to migrate their networks to SDN enabled IPv6 Network. Additionally, SDN provides better visibility of the network and IPv6 supports a large number of devices to identify uniquely in the global scale. In this paper, we first established the economic model of Tier-3 ISPs during their network migration and then demonstrated the evolutionary process with constraints on joint migration so that ISPs can take their own decision for migration based on their adaptation strength considering customer's demands, interactions with other ISPs and status of organizational capital and operational expenditure. Our study shows that migrating traditional network into SoDIP6 network is affected by different internal and external parameters in which migration decision when taken in the proper time step could result optimum organizational cost for service providers and get higher utilities.

Acknowledgment. This research is supported by University Grant Commission (UGC), Nepal (grant id: FRG/74_75/Engg-1) and NTNU (Norwegian University of Science and Technology) under Sustainable Engineering Education Project (SEEP). We also received partial funding support from Nepal Academy of Science and Technology (NAST). Additionally, the work of Dr. Danda B Rawat was partly supported by the U.S. National Science Foundation (NSF) under grants CNS 1650831 and HRD 1828811. However, any opinion, finding, and conclusions or recommendations expressed in this document are those of the authors and should not be interpreted as necessarily representing the official policies, either expressed or implied, of the NSF.

References

1. Dawadi, B.R., Rawat, D.B., Joshi, S.R.: Software defined IPv6 networks: a new paradigm for future networks. J. Inst. Eng. **15**(2), 1–14 (2018)
2. Dawadi, B.R., Joshi, S.R., Khanal, A.R.: Service provider IPv4 to IPv6 network migration strategies. J. Emerg. Trends Comput. Inf. Sci. **6**(10), 565–572 (2015)
3. Wu, P., Cui, Y., Wu, J., Liu, J., Metz, C.: Transition from IPv4 to IPv6: a state-of-the-art survey. IEEE Commun. Surv. Tutor. **15**(3), 1407–1424 (2013)
4. Dawadi, B.R., Rawat, D.B., Joshi, S.R., Keitsch, M.M.: Joint cost estimation approach for service provider legacy network migration to unified software defined IPv6 network. In: IEEE 4th International Conference on Collaboration and Internet Computing (CIC), Philadelphia/PA/USA, pp. 372–379 (2018). https://doi.org/10.1109/cic.2018.00056
5. Rizvi, S.N., Raumer, D., Wohlfart, F., Carle, G.: Towards carrier grade SDNs. Comput. Netw. **92**, 218–226 (2015)
6. ON.LAB Whitepaper, Introducing ONOS - a SDN network operating system for Service Providers (2014). http://onosproject.org/wp-content/uploads/2014/11/Whitepaper-ONOS-final.pdf. Accessed Oct 2018
7. Karakus, M., Durresi, A.: Economic viability of software defined networking (SDN). Comput. Netw. **135**, 81–95 (2018)

8. Voelkl, B.: Simulation of Evolutionary Dynamics in Finite Populations. www.mathematica-journal.com/data/uploads/2011/05/Voelkl.pdf. Accessed Nov 2018
9. Nowak, M.A.: Evolutionary Dynamics. Harvard University Press (2006)
10. Trinh, T.A., Gyarmati, L., Sallai, G.: Migrating to IPv6: a game-theoretic perspective. In: IEEE 35th Conference Local Computer Networks (LCN), pp. 344–347 (2010)
11. Nikkhah, M.: Maintaining the progress of IPv6 adoption. Comput. Netw. **102**, 50–69 (2016)

Hiding Patient Injury Information in Medical Images with QR Code

Akkarat Boonyapalanant, Mahasak Ketcham[(✉)],
and Manussawee Piyaneeranart

Faculty of Information Technology,
ITMRC, King Mongkut's University of Technology North Bangkok,
Bangkok, Thailand
{akkarat.b,Mahasak.k}@it.kmutnb.ac.th,
s5907011910045@email.kmutnb.ac.th

Abstract. Watermarking has been applied in the medical field that is used to enhance the safety of medical information. QR Code is used in this research to store medical image data and insert a watermark into the image using the Least Significant Bit - LSB method that can insert data into the bit sensitive area. Watermark insertion using the LSB method does not affect the image size and cannot be seen by the eye. This method insert a watermark that is distributed throughout the image. The experimental have rotated the image in 90 degrees in a clockwise direction, rotated 90 degrees in a counterclockwise direction and rotated in the opposite direction. The results of the experiment showed that the rotation of the image in the above direction did not affect the reading of the patient's injury data from the QR Code.

Keywords: QR code · Watermarking · Medical image

1 Introduction

Steganography is a method that has been used since 440 BC which different forms [1]. Nowadays, the method is used in many forms by applying to images that are on social media. Images must be compressed before storage and encrypted for image safety. Encryption [2] hides only the meaning or content of the message from the eavesdropper, but the encrypted message remains and visible. Steganography is the concept of inserting information that cannot be verified by the human senses into the visual, audio, and video media. This technique is called digital watermarking [3]. Data hiding techniques [4] are in the form of camouflage or steganography, which is inserted into digital media. The purpose of hiding information is to identify annotation and in the matter of copyright. The data hiding process has many limitations, including the quality of hidden data of loss of compressed data. There is used to enhance the safety of medical information by introducing zero watermarking [5] to solve security problems and reduce the loss of clinical information that adopting Singular Value decomposition: SVD to create digital binary code for each image. The researcher has introduced the technique of zero watermarking developed as Robust Zero-watermarking and work with Perceptual Hashing [6] to provide a robust watermark for attack and have researchers have

P. Boonyopakorn et al. (Eds.): IC2IT 2019, AISC 936, pp. 258–267, 2020.
https://doi.org/10.1007/978-3-030-19861-9_25

developed zero watermarking techniques that are safer and more resistant to attack patterns. The technique is called A New Zero-watermarking by working with the hashing algorithm [7]. In addition, there are separate wavelet transform technique: DWT and Singular Value decomposition: SVD [8] has been applied to medical imaging such as X-ray, CT Scan and mammography. This technique can control privacy, Electronic Patient Record (EPR), easily retrieved and reduce data loss. The researcher has adopted a discrete wavelet transform technique: DWT and Singular Value decomposition: SVD for medical imaging, including imaging [9], with a watermark encoded for increased durability even though attack, but does not affect the quality of the watermark image. Watermark insertion techniques is used to store data online and offline with QR codes that use the Wavelet Based Annotation [10]. Wavelet transform technique (DWT) with QR code [11] is used to hide outpatient information about medical images in dialysis using Level 3 wavelet transform with QR code which is resistant to multi attacks. When users want to send information to be safe [12] inserting large secret data may involve inserting the QR codes into the RGB channel. In addition, new methods can be used in reverse watermark reading [13], which uses a binary format as a watermark. From above, medical storage has a problem in losing data about treatment methods and clinical data is too large to be inserted into the image. This article presents the QR code used to store medical data and insert watermarks into images using the Least Significant Bit – LSB, ensuring that the clinical data of the image will not be lost.

2   Theory

2.1   Hiding and Steganography

Data hiding and data embedding [14] typically refers to an application with data inserted in various media. The basic technical definitions are as follows:

(1) Data Hiding: is a visual challenge for the human visual system (HVS) and a visual modification [3]. Still images are changed from normal to non-linear such as cropping, blurring, filtering, and compression. Useful hiding techniques need to be as robust as possible to convert these data as possible.
(2) Steganography [15] is the art of hiding information in a format that prevents text messages from being detected. Steganography from Greek language, there is a way to communicate in secret that is hidden in the message. The encoding makes the message incomprehensible for camouflage is to hide the message to not see.

2.2   Watermarking

Watermark is the process of inserting special data into a media such as an image, sound, and video which can be retrieved from the multimedia content at a later time and used to support ownership.

(1) Type of digital watermarking. [16]
    (a) Visible Watermarking: The purpose of this digital watermark to show ownership of the work, such as corporate stamps or the logo of the TV station in

the lower right corner of the television screen make the viewer aware that the information belongs to any person or entity.

(b) Invisible watermarking: This type requires an encryption code encrypt the signal to prevent unauthorized changes from unauthorized persons, which only the owner knows the key encryption. Therefore, people cannot know what is inserted in the data, even if the person knows the insertion and deletion mechanism of the watermark signal.

(2) Digital watermark insertion technique. [17]

(a) Robust Watermarking Techniques: This technique is particularly well suited for applications that relate to piracy prevention because of the difficulty of altering, modifying, or damaging watermark signals.

(b) Fragile Watermarking Techniques: Fragile watermarking is a sensitive and easy way to modified data. This type of technique is best suited to the credibility of the data that is genuine without any modifications.

2.3   Quick Response (QR Code)

QR Code is a 2D bar code matrix. There is similarity of a square barcode, can contain as many as 7,089 numbers or 4,296 characters, binary number data, 2,953 bytes and 1,817 Japanese characters. Contains Black modules are Binary 1 and the QR code has tree corner finder patterns. The three finder patterns are used for positioning in which QR codes can store up to 7,089 characters of Numeric, Alphanumeric, Kanji, Kana, Hiragana, Symbols, Binary, and control codes 1 image of QR code [17, 18]. All of this shown in Fig. 1.

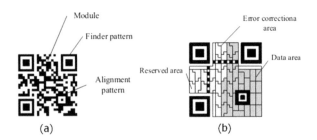

Fig. 1. Structure of QR code.

The structure of the QR Code has three corners. The three corners of the QR code are the left, top, bottom, and upper right corner (finder patterns) and the alignment patterns for fast reading. The data is encoded and checked for errors (data and correction code words) and the storage area of the error information (format information) according to the image shown in this Fig. 2 [12].

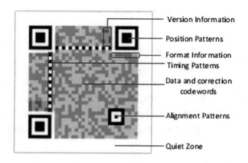

Fig. 2. Display the structure of the QR code

2.4 Steganography Using Inserting the Least Significant Bit (LSB)

Least Significant Bit - LSB is the last bit changing method. Mostly, the color value used for uncompressed image files is 3 bytes in 1 pixel color.

The most commonly used color for uncompressed image files is 3 bytes for 1 pixel color. The 3-byte color value is Red, Green and Blue in one byte. Can be represented the letter as R, G, B, respectively as shown in Fig. 3.

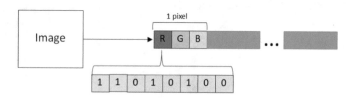

Fig. 3. Display pixel RGB elements.

A pixel consists of R, G, B that stand for Red, Green, and Blue, respectively. Each value contains an 8-bit binary value equivalent to one pixel, which is a total of 24 bits. So when using images as a medium for inserting data into an image, there are ways to do as shown in Fig. 4.

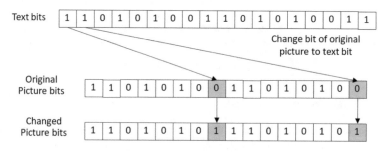

Fig. 4. Steganography on image, pixel form.

3   Methodology

This research, the QR code is used to store clinical information of medical images and then insert into the image by Least Significant Bit – LSB which maintains the medical image data not to be lost for use with long distance therapy systems. This method inserts a QR code into a medical image, so that cannot be seen by the human eye. This method is called Invisible Watermarking. In this research we used QR code as the encryption key to disguise the data. The implementation process is shown in Fig. 5.

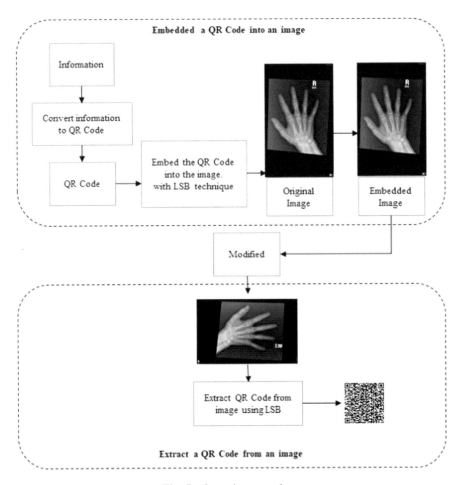

Fig. 5. Operating procedures

3.1   Text Conversion to QR Code

Text to be stored in the QR code as shown in Fig. 6(a). Bring text into QR code conversion process as shown in Fig. 6(b).

Demographics
13 y.o. boy

Caption
Right hand 9-28-05 : AP view
demonstrates several small osteosclerotic
lesions with most in a periarticular pattern.
Broken needle tip adjacent to the third
finger terminal tuft.

(a)   (b)

Fig. 6. Text conversion to QR code

First data analysis, input data is a number use numeric, input data is all characters use alphanumeric mode, input data is UTF-8 use byte mode, input data is a JIS character to use Kanji mode. Second data encoding generates code words to help correct read errors. Third error correcting code words, QR code reader reads both code words data and code words error corrections. When the QR code reader is analyzed that compares whether the data is read correctly and the error can be corrected if the data cannot be read. Finally code words, data alignment and error correction code words created in the previous step are correct. Then place the bits in the QR code's matrix.

3.2 Image Preparation

In this research, the image used in the experiment is a BMP file of about 300×300 pixels. Due to the large size of the image used in the experiment this will cause processing errors resulting in incomplete QR code insertion. In the case of large images will be scaled down as shown in Fig. 7.

Fig. 7. Image reduction

3.3 Inserting and Recovery QR Code

(1) Inserting QR Code into image

The picture, as shown in Fig. 8(a), has been inserted QR code with a size of 100×100 pixels, as shown in Fig. 8(b). If we used a QR code larger than 100×100 pixels and QR code inserted into the image, then when remove the QR code will produce a QR Code in case the image is changed the QR code will be destroyed and cannot read from QR Code. If the QR Code smaller than 100×100 pixels when decoding the QR code will not be able to read the data properly. Therefore, use the QR code of 100×100

pixels, which is suitable for insertion image when decoding the QR code and the size is appropriate, which can be easily read and if the image has changed the QR code, some code read the QR code shown in Fig. 8(b).

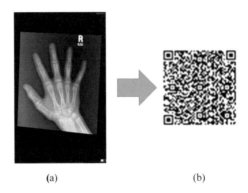

(a)   (b)

Fig. 8. Original image and QR code image

(2) Steganography by LSB method

The QR code is inserted into the image this method is disguised on the RGB channel, the last bit of the channel is inserted into the bit with the least significant value as show in Fig. 9.

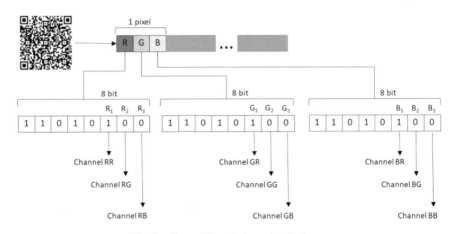

Fig. 9. Show QR code insertion in image

(3) QR code recovery from images

Inserting QR codes to detect changes in QR codes using the Least Important Bit - LSB method. In decoding QR, there is a step opposite the encoding process. In this research, we use the QR Code reader program to read QR codes.

4 Experimental and Result

This research insert QR Code into the color image which the encryption the signal. The color image is obtained through the insertion of a watermark after that has been modified. For checking the QR Code decoded to verify that the QR Code is complete or destroyed. From Fig. 10, the original image has been inserted into the QR code and when decoding the QR code, the content can be read into the QR code (Figs. 11, 12 and 13).

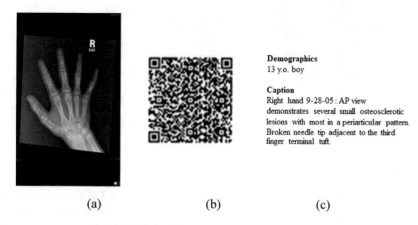

(a)   (b)   (c)

Fig. 10. Original image has been inserted QR code.

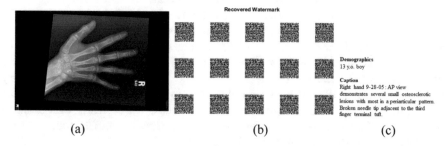

(a)   (b)   (c)

Fig. 11. The image rotates at an angle of 90 degrees in the clockwise direction.

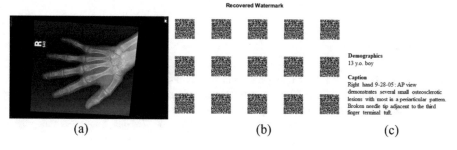

(a)   (b)   (c)

Fig. 12. The image rotates at an angle of 90 degrees in the counterclockwise direction.

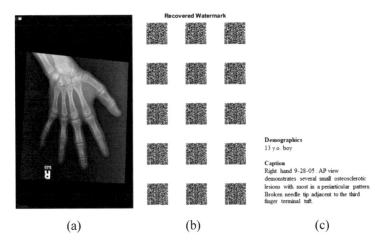

(a) (b) (c)

Fig. 13. The image rotates in the opposite direction to the original image.

5 Conclusion

This research has experimented with inserting watermark using QR Code inserted into image by Least Significant Bit - LSB method. This watermark insert does not make the image file larger and this watermark insert is not visible to the human eye. This corresponds to the purpose of this watermark insertion that requires medical image data collection to prevent data loss using the Least Significant Bit - LSB method.

The experiment shown in part IV, we have attacked the image by rotating the image. The image has been rotated in a 90-degree clockwise direction, the 90-degree counterclockwise direction, and the opposite direction. After the image has been attacked by rotating the image, we have decoded the QR. From the experimental, we found that the QR code of the image was rotated when decoded did not find any damage because the QR code has technical features to put a robust watermark. We read data from QR codes with a program to read QR codes. We can read medical information from QR codes. Medical information has been read from QR codes that are complete and not lost from image attacks by rotation.

Acknowledgment. This research was funded by King Mongkut's University of Technology North Bangkok, Contract no. KMUTNB-60-GEN-018.

References

1. Gupta, R., Jain, S.: A review on watermarking techniques for compressed encrypted images. Presented at the International Conference Greater Medical Imaging, m-Health and Emerging Communication Systems (MedCom), Noida, 7–8 November 2014
2. Paliwal, S., Nigam, R.K.: A Survey & Applications of Various Image Steganography Techniques

3. Lee, S.-J., Sung-Hwan, J.: A survey of watermarking techniques applied to multimedia. Presented at the International Symposium on Industrial Electronics Proceedings (Cat. No. 01TH) 8570 ISIE 2001. IEEE, 12–16 June 2001
4. Bender, W., Gruhl, D., Morimoto, N., Lu, A.: Techniques for data hiding. IBM Syst. J. **35** (3.4), 313–336 (1996)
5. Singh, A., Raghuvanshi, N., Dutta, M.K., Burget, R., Masek, J.: An SVD based zero watermarking scheme for authentication of medical images for tele-medicine applications. Presented at the 39th International Conference on Telecommunications and Signal Processing (TSP), 27–29 June 2016
6. Baoru, H., Jingbing, L., Mengxing, H.: A robust zero-watermarking algorithm against geometric attacks with perceptual hashing. Presented at the Metallurgical and Mining Industry (2015). Article
7. Han, B., Li, J.: A new zero-watermarking algorithm resisting attacks based on differences hashing. Presented at the Cybernetics and Information Technologies (2016)
8. Thakkar, F.N., Srivastava, V.K.: A blind medical image watermarking: DWT-SVD based robust and secure approach for telemedicine applications. Presented at the Multimedia Tools and Applications, 01 February 2017. journal article
9. Singh, A.K., Kumar, B., Dave, M., Mohan, A.: Robust and imperceptible dual watermarking for telemedicine applications. Presented at the Wireless Personal Communications, 01 February journal article (2015)
10. Kazakeviciute-Januškeviciene, G., Ušinskas, A., Januškevicius, E., Ušinskiene, J., Letautiene, S.: Annotation of the medical images using quick response codes. Presented at the Computer Science and Information Systems (2017)
11. Sriyapai, C.: "Watermarking Approach of Hiding Patient Information with QR code for Authentication in Telemedicine," Master, Management Information Systems. King Mongkut's University of Technology North, Bangkok (2015)
12. Dangmee, P., Leelaketsakul, W.: Steganography hiding data within QRCode. J. Inf. Sci. Technol. **5**(1) (2015)
13. Hsu, F.-H., Wu, M.H., Wang, S.J.: Dual-watermarking by QR-code applications in image processing. Presented at the Ubiquitous Intelligence and Computing and 9th International Conference on Autonomic and Trusted Computing, 4–7 September 2012
14. Hartung, F., Kutter, M.: Multimedia watermarking techniques. Presented at the Proceedings of the IEEE (1999)
15. Johnson, N.F., Jajodia, S.: Exploring steganography: Seeing the unseen. Presented at the Computer (1998)
16. Panyavaraporn, J.: Digital watermark in image and video. Burapha Univ. J. **19**(2), 210–226 (2014)
17. Chang, Y.-H., Chu, C.H., Chen, M.S.: A general scheme for extracting QR code from a. In: 2007 Ninth IEEE International Symposium Multimedia (ISM), pp. 123–130 (2007)
18. D.W. Incorporated. About QR Code (1994). http://www.denso-wave.com/qrcode/index-e. html

Author Index

A
Alphonse, P. J. A., 213
Atsawaruangsuk, Sarutte, 148

B
Bheganan, Poramin, 69, 99
Boonsiri, Somjai, 37
Boonyapalanant, Akkarat, 258

C
Chiracharit, Werapon, 46
Chompoopuen, Hathaichanok, 129
Chupraphawan, Supakorn, 169

D
Dawadi, Babu R., 245
Doan, Van Thang, 227

H
Hemmi, Kazuo, 119
Huynh, Cong Phap, 227

I
Imamura, Kosuke, 205

J
Jaupunphol, Pita, 129
Jitkajornwanich, Kulsawasd, 26
Joshi, Shashidhar R., 245

K
Kachai, Tontrakant, 69, 99
Kanchanapreechakorn, Sarattha, 111
Katanyukul, Tatpong, 148
Kee, Kerk F., 26

Ketcham, Mahasak, 258
Khang, Vo Quang Hoang, 227
Kittiphattanabawon, Nichnan, 59
Klinkasen, Pokpakorn, 159
Kondo, Yuki, 119
Kumar, Nitin, 79
Kurdthongmee, Wattanapong, 138
Kusakunniran, Worapan, 111

L
Laopracha, Natthariya, 180
Lawawirojwong, Siam, 26
Lekpong, Peerapon, 26
Lertrungwichean, Khitichai, 159
Li, Jiangfeng, 15
Luaphol, Bancha, 69

M
Meesad, Phayung, 227
Mittrapiyanurak, Pradit, 46

N
Nasomboon, Kanyarat, 99
Netisopakul, Ponrudee, 89
Nguyen, Ha Huy Cuong, 227
Nishimura, Takeshi, 119
Nuchprasert, Chawanwut, 159

P
Pasupuleti, Syam Kumar, 213
Petrasch, Roland, 235
Phetkrachang, Ketsara, 59
Piyaneeranart, Manussawee, 258
Polpinij, Jantima, 69, 99
Polpinit, Pattarawit, 148

© Springer Nature Switzerland AG 2020
P. Boonyopakorn et al. (Eds.): IC2IT 2019, AISC 936, pp. 269–270, 2020.
https://doi.org/10.1007/978-3-030-19861-9

Pongto, Ratchanont, 26
Premkamal, Praveen Kumar, 213
Pumruckthum, Pollaton, 37

R
Ratanamahatana, Chotirat Ann, 3, 169
Rawat, Danda B., 245

S
Shi, Yang, 15
Simcharoen, Supaporn, 193
Sinapiromsaran, Krung, 37
Srikanjanapert, Natthakit, 69
Srikudkao, Boonchoo, 69
Srisonphan, Siwapon, 26

T
Techawatcharapaikul, Chanchai, 46

Thong-iad, Kanlaya, 89
Thongkanchorn, Kittikhun, 111
Thumthong, Wichan, 129
Tobina, Takuro, 119

U
Udomchaiporn, Akadej, 159

V
Vichit, Nattakit, 3

W
Wiwattanaphon, Nopparat, 26

Z
Zhang, Yinjia, 15
Zhao, Qinpei, 15

Printed in the United States
By Bookmasters